计算机类技能型理实一体化新形态系列

Java Web 程序设计

（微课视频版）

主　编　吴绍根　张寺宁

清华大学出版社
北京

内容简介

本书是一本介绍 Java Web 程序设计的基础图书,知识内容与 Servlet 6.0 版技术规范同步,适合 Java Web 初学者使用。全书共有三大部分,包括 12 章,全面介绍了使用 Java Web 技术开发 Web 应用程序的基本概念、基本方法、基本技术,各章均配有综合案例,对知识内容进行总结性应用。第 1 部分为 Java Web 核心技术,包括第 1~7 章,介绍了 Java Web 的关键核心技术,具体包括:建立 Java Web 开发环境、HTTP 传输协议、Servlet 关键技术、会话管理、数据库连接池及其使用方法等;第 2 部分为 Java Web 表示技术,包括第 8~11 章,介绍了 Java Web 数据表示相关技术,具体包括:系统分层结构及 MVC 设计模式、JSP 表示技术、Thymeleaf 表示技术、JSON、JavaScript 和 Ajax;第 3 部分为 Java Web 高级特性,只包括第 12 章,介绍了 Servlet 的 AsyncContext 异步处理请求技术和 Non Blocking I/O 技术。

本书可作为高校计算机相关专业的教材,也可作为 Java Web 编程爱好者的自学书籍。

本书封面贴有清华大学出版社防伪标签,无标签者不得销售。
版权所有,侵权必究。举报: 010-62782989,beiqinquan@tup.tsinghua.edu.cn。

图书在版编目(CIP)数据

Java Web 程序设计:微课视频版/吴绍根,张寺宁
主编. --北京:清华大学出版社,2024.9. --(计算
机类技能型理实一体化新形态系列). --ISBN 978-7
-302-67177-0

Ⅰ.TP312.8

中国国家版本馆 CIP 数据核字第 2024MU1005 号

责任编辑:张龙卿
封面设计:刘代书 陈昊靓
责任校对:李 梅
责任印制:沈 露

出版发行:清华大学出版社
网   址:https://www.tup.com.cn,https://www.wqxuetang.com
地   址:北京清华大学学研大厦 A 座   邮   编:100084
社 总 机:010-83470000   邮   购:010-62786544
投稿与读者服务:010-62776969,c-service@tup.tsinghua.edu.cn
质量反馈:010-62772015,zhiliang@tup.tsinghua.edu.cn
课件下载:https://www.tup.com.cn,010-83470410

印 装 者:三河市天利华印刷装订有限公司
经   销:全国新华书店
开   本:185mm×260mm   印   张:18.75   字   数:454 千字
版   次:2024 年 9 月第 1 版   印   次:2024 年 9 月第 1 次印刷
定   价:59.80 元

产品编号:102067-01

前　言

　　Java Web 是 Java EE 的核心组成部分，结合 Java Web 技术和流行的框架技术，可以开发多种类型的网络应用系统。

　　Java Web 是 Java 体系的重要组成部分。Java 在体系结构上包括以下几个层次：Java SE（Java Standard Edition，Java 核心）、Java EE（Java Enterprise Edition，Java Web 是 Java EE 中的一个主要部分）、Java 框架（Java Spring Framework）、Java Micro Service（Java 微服务）和 Java Cloud（Java 云端开发部署）。其中，基于 Java 核心可以开发 Java 桌面应用程序；结合 Java Web 技术、Java 框架技术、Java Micro Service 技术和 Java Cloud 技术可以开发各种规模的因特网应用系统，例如，目前的一些典型网络应用系统包括电子政务系统、电子商务系统等都是基于这些技术体系开发和建设的。

　　为了帮助学习者快速掌握和使用 Java 技术开发应用程序，清华大学出版社携手高校和企业有经验的教师、工程师开发了一整套 Java 技术体系丛书，本丛书共 5 本，包括《Java 面向对象程序设计（微课视频版）》《Java Web 程序设计（微课视频版）》《Spring 框架应用开发——基于 Spring Boot（微课视频版）》《Spring Cloud 微服务应用开发——基于 Alibaba Nacos（微课视频版）》《Spring 微服务系统部署（微课视频版）》。

　　本书共有三大部分，包括 12 章，介绍了 Java Web 的关键核心基础知识和相关技术。

　　学习 Java Web 的目的是在工程项目中使用 Java Web 开发应用系统。本书的典型特点是除了对 Java Web 重要知识和技能进行通俗易懂的介绍外，在每章均安排了"综合案例"，通过这些案例介绍各章所述知识点的具体应用，同时，对重要的知识点还安排了"最佳实践"，以便能够学以致用。每章均配有课后练习题。

　　第 1 章介绍如何建立 Java Web 开发环境。本书采用目前较为流行的最新版 IntelliJ IDEA 高级版作为 Java Web 开发环境。第 2 章介绍 Java Web 技术的系统模型，重点介绍了 HTTP（超文本传输协议）的应用，包括 HTTP 请求和 HTTP 响应，并对如何跟踪 HTTP 应用过程做了介绍。第 3 章介绍 Servlet 的核心技术内容及其相关接口、类的使用。第 4 章介绍如何使用 Servlet 技术进行文件的上传、图片文件的下载及显示、普通文件的下载等技术。第 5 章介绍用于管理和维护会话的常用技术,重点介绍了 HttpSession 会话管理。第 6 章介绍监听器和过滤器的使用。第 7 章介绍如何在 Java Web 程序中访问数据库数据,重点介绍了使用 JDBC 访问数据库数据。第 8 章介绍在设计系统时分层结构和编码的必要性,重点介绍了 MVC 设计模式。第 9 章对 JSP 技术进行了仔细梳理,重点介绍了 JSP 技术的 EL 表达式和 JSTL 标签技术,使数据表示简洁明了,并且代码具有较好的可读性和可维护性。第 10 章介绍如何用 Thymeleaf 引擎展示 Web 数据。Thymeleaf 作为 Spring 推荐的表示技术,是一种高效优雅的数据表示技术。第 11 章介绍 JSON、JavaScript 和 Ajax 相关知识和技术。第 12 章介绍 Servlet 的高级特性,包括 AsyncContext 异步处理请求技术和 Non Blocking I/O 技术,使用这些新技术可以有效提高系统的执行效率。

  本书建议授课 72 课时左右,各院校也可根据具体情况适当调整。

  本书的第 1、2 章由张寺宁编写,第 3~12 章由吴绍根编写。本书配有详细的 PPT 讲义、教学视频、书本源代码、课后练习解答等电子资源,这些电子资源可从清华大学出版社官网下载或扫码观看。

编 者
2024 年 4 月

目 录

第1部分 Java Web 核心技术

第1章 建立 Java Web 开发环境 3
- 1.1 Java Web 概述 3
  - 1.1.1 前端服务程序 3
  - 1.1.2 后端服务程序 4
- 1.2 建立 Java Web 开发环境 4
- 1.3 开发第一个 Java Web 程序 5
  - 1.3.1 新建 Java Web 项目 5
  - 1.3.2 编写登录页面代码 7
  - 1.3.3 运行和访问登录页面 10
  - 1.3.4 IDEA 中或页面中出现乱码的解决方法 12
- 1.4 C/S 架构和 B/S 架构 12
- 1.5 练习：建立 Java Web 开发环境 13

第2章 HTTP 超文本传输协议 14
- 2.1 前端服务程序与后端服务程序之间的通信 14
  - 2.1.1 HTTP 通信模型 14
  - 2.1.2 HTTP 通信过程跟踪 15
- 2.2 HTTP 请求消息 18
  - 2.2.1 认识 URL 18
  - 2.2.2 HTTP 请求消息格式 19
  - 2.2.3 HTTP 请求方法 19
  - 2.2.4 HTTP 常见请求头 20
- 2.3 HTTP 响应消息 21
  - 2.3.1 HTTP 响应消息格式 21
  - 2.3.2 HTTP 状态码 22
  - 2.3.3 HTTP 常见响应头 23
  - 2.3.4 响应体 24

2.4 对HTTP请求进行深入跟踪剖析 …… 24
2.5 练习：跟踪浏览器请求和服务器的响应 …… 26

第3章 Servlet基础 …… 27

3.1 Servlet入门 …… 27
3.1.1 完善登录页面 …… 27
3.1.2 创建ch03工程 …… 28
3.1.3 创建Login后端服务程序 …… 29

3.2 Servlet的具体应用 …… 31
3.2.1 登录页面与后端服务程序Login的交互过程 …… 31
3.2.2 @WebServlet注解 …… 36
3.2.3 Servlet接口及生命周期 …… 37
3.2.4 Servlet接口的实现类GenericServlet和HttpServlet …… 39

3.3 案例：更为完整的Login Servlet程序 …… 40
3.3.1 案例目标 …… 40
3.3.2 案例分析 …… 41
3.3.3 案例实施 …… 41

3.4 HttpServletRequest对象及其使用 …… 45
3.4.1 获取请求参数 …… 46
3.4.2 获取HTTP请求头信息 …… 47
3.4.3 转发请求和页面包含 …… 49

3.5 HttpServletResponse对象及其应用 …… 52
3.5.1 发送响应数据到客户端 …… 52
3.5.2 深入了解setContentType方法和MIME …… 54
3.5.3 发送状态码或错误信息 …… 55
3.5.4 设置响应消息头信息 …… 57
3.5.5 请求重定向 …… 59

3.6 ServletContext对象及其使用 …… 60
3.6.1 使用ServletContext实现数据共享 …… 61
3.6.2 使用ServletContext读取资源文件 …… 64
3.6.3 关于web.xml配置文件 …… 66

3.7 案例：用户注册 …… 67
3.7.1 案例目标 …… 67
3.7.2 案例分析 …… 67
3.7.3 案例实施 …… 67

3.8 练习：编写书籍录入程序 …… 71

第4章 Servlet文件上传和下载 …… 72

4.1 Servlet接收上传文件 …… 72

		4.1.1	编写包含上传文件功能的注册页面	73
		4.1.2	接收客户端上传的头像文件	74
		4.1.3	多文件上传	77
	4.2	Servlet 下载文件到客户端		79
		4.2.1	下载并显示图像	80
		4.2.2	下载并保存图像文件	82
		4.2.3	下载和保存任意类型的文件	83
	4.3	案例：美图分享		85
		4.3.1	案例目标	86
		4.3.2	案例分析	86
		4.3.3	案例实施	87
	4.4	练习：完善书籍录入程序		91

第 5 章 会话管理 … 92

	5.1	会话及其常用技术		92
	5.2	Cookie 技术		93
		5.2.1	什么是 Cookie	93
		5.2.2	Cookie 类	93
		5.2.3	使用 Cookie 实现会话管理举例	94
		5.2.4	Cookie 观察和 Cookie 使用注意事项	99
	5.3	Session 技术		101
		5.3.1	HttpSession 接口	101
		5.3.2	使用 HttpSession 管理会话举例	102
		5.3.3	Session 观察	104
	5.4	案例：简单的购物系统		106
		5.4.1	案例目标	106
		5.4.2	案例分析	106
		5.4.3	案例实施	107
	5.5	练习：记录用户上次登录的时间和地点		112

第 6 章 Servlet 监听器和过滤器 … 113

	6.1	Servlet 监听器		113
		6.1.1	监听 ServletContext 对象	113
		6.1.2	监听 HttpSession 对象	118
		6.1.3	监听 HttpServletRequest 对象	119
	6.2	Filter 过滤器		120
		6.2.1	Filter 接口及其实现类 HttpFilter	121
		6.2.2	Servlet 过滤器应用举例	122
		6.2.3	FilterChain 接口	125

6.3 案例：使用过滤器检查用户登录状态 ··· 125
  6.3.1 案例目标 ··· 125
  6.3.2 案例分析 ··· 126
  6.3.3 案例实施 ··· 126
6.4 练习：选班长 ··· 132

第7章 访问数据库 ··· 133

7.1 使用 JDBC 访问数据库 ·· 133
  7.1.1 使用 JDBC 访问数据库的一般过程 ·· 133
  7.1.2 使用 JDBC 访问数据库示例 ··· 133
7.2 数据库连接池 ··· 138
  7.2.1 什么是数据库连接池 ·· 138
  7.2.2 DataSource 接口 ·· 139
  7.2.3 使用 DBCP 建立数据库连接池 ·· 139
  7.2.4 使用 Druid 建立数据库连接池 ··· 142
7.3 案例：将用户注册信息保存到数据库 ··· 146
  7.3.1 案例目标 ··· 146
  7.3.2 案例分析 ··· 146
  7.3.3 案例实施 ··· 147
7.4 练习：将图书信息保存到数据库 ·· 155

第2部分 Java Web 表示技术

第8章 系统分层结构及 MVC 设计模式 ·· 159

8.1 程序功能部件之间的耦合度 ··· 159
8.2 Java Web 程序的分层结构 ·· 159
8.3 Java Web 的 MVC 设计模式 ··· 160
8.4 常用的 Java Web 表示技术 ·· 161

第9章 JSP 表示技术 ·· 163

9.1 JSP 作为 MVC 的表示技术 ··· 163
  9.1.1 第一个 JSP 程序 ··· 163
  9.1.2 JSP 的工作原理 ··· 165
9.2 JSP 程序组成 ··· 168
  9.2.1 JSP 指令 ··· 168
  9.2.2 JSP 脚本 ··· 169
9.3 EL 表达式 ·· 173
  9.3.1 EL 表达式基本语法及 EL 表达式内置对象 ······································· 173
  9.3.2 EL 表达式运算符 ··· 177

9.4 JSTL 标签及其使用 179
 9.4.1 如何使用 JSTL 标签库 179
 9.4.2 JSTL 核心标签 180
9.5 JSP 最佳实践 190
9.6 案例：图书信息管理系统 191
 9.6.1 案例目标 191
 9.6.2 案例分析 191
 9.6.3 案例实施 192
9.7 练习：学生信息管理系统 206

第 10 章 Thymeleaf 表示技术 207

10.1 Thymeleaf 作为 MVC 表示技术 207
 10.1.1 导入 Thymeleaf 到项目工程 207
 10.1.2 创建 Thymeleaf 引擎 208
 10.1.3 使用 Thymeleaf 引擎生成结果页面 210
10.2 Thymeleaf 模板表达式 212
 10.2.1 消息表达式♯{...} 212
 10.2.2 变量表达式 ${...} 214
 10.2.3 选择对象表达式 *{...} 218
 10.2.4 URL 链接表达式@{...} 220
10.3 Thymeleaf 的字面常量和运算符 223
 10.3.1 字面常量 223
 10.3.2 字符串操作 223
 10.3.3 算术运算、关系运算和逻辑运算 224
 10.3.4 条件运算符 224
 10.3.5 字面常量和运算符使用举例 224
10.4 Thymeleaf 常用属性及其使用 226
 10.4.1 使用 th:text、th:utext 和内联属性输出文字 226
 10.4.2 使用 th:with 属性定义局部变量 226
 10.4.3 使用 th:attr 属性设置 HTML 标签的属性值 227
10.5 Thymeleaf 的条件控制和迭代 228
 10.5.1 th:each 迭代的使用 228
 10.5.2 th:if 和 th:unless 条件控制的使用 231
 10.5.3 th:switch/th:case 多分支控制的使用 232
10.6 Thymeleaf 工具类及其使用 232
10.7 案例：图书信息管理系统 235
 10.7.1 案例目标 235
 10.7.2 案例分析 235
 10.7.3 案例实施 236

10.8 练习：学生信息管理系统 ………………………………………………… 247

第 11 章　JSON、JavaScript 和 Ajax …………………………………………… 248

11.1 JSON 及其使用 ………………………………………………………… 248
　　11.1.1 JSON 基础 ……………………………………………………… 248
　　11.1.2 为什么需要 JSON ……………………………………………… 248
　　11.1.3 在 Servlet 程序中处理 JSON 数据 …………………………… 250
11.2 JavaScript 和 Ajax ……………………………………………………… 253
　　11.2.1 展示所有书籍信息 ……………………………………………… 253
　　11.2.2 Ajax ……………………………………………………………… 255
11.3 案例：图书信息管理系统 ……………………………………………… 260
　　11.3.1 案例目标 ………………………………………………………… 260
　　11.3.2 案例分析 ………………………………………………………… 260
　　11.3.3 案例实施 ………………………………………………………… 260
11.4 练习：完善图书信息管理系统 ………………………………………… 273

第 3 部分　Java Web 高级特性

第 12 章　Servlet 高级技术 ……………………………………………………… 277

12.1 AsyncContext 异步处理请求技术 ……………………………………… 277
　　12.1.1 AsyncContext 入门示例 ………………………………………… 277
　　12.1.2 AsyncContext 接口 ……………………………………………… 280
　　12.1.3 AsyncListener 监听器接口 ……………………………………… 280
12.2 Non Blocking I/O 技术 ………………………………………………… 283
12.3 案例：使用 AsyncContext 访问第三方系统 ………………………… 284
　　12.3.1 案例目标 ………………………………………………………… 284
　　12.3.2 案例分析 ………………………………………………………… 284
　　12.3.3 案例实施 ………………………………………………………… 286
12.4 练习：使用 Thymeleaf 显示气象数据 ………………………………… 288

参考文献 …………………………………………………………………………… 289

Java Web 核心技术

- 第 1 章　建立 Java Web 开发环境
- 第 2 章　HTTP 超文本传输协议
- 第 3 章　Servlet 基础
- 第 4 章　Servlet 文件上传和下载
- 第 5 章　会话管理
- 第 6 章　Servlet 监听器和过滤器
- 第 7 章　访问数据库

第 1 章　建立 Java Web 开发环境

Java 技术体系从技术层次上包括两个部分：第一部分是 Java SE，也就是 Java Standard Edition，称为 Java 标准版，是 Java 技术的基础内容。使用 Java SE 可以基于 Java 面向对象技术、方法和基础类库开发桌面应用程序；第二部分是 Java EE，也就是 Java Enterprise Edition，称为 Java 企业版，是 Java 技术的扩展内容。使用 Java EE 可以开发各种类型的网络应用程序，如电子商务系统、电子政务系统等。Java Web 是 Java EE 的核心部分，使用 Java Web 可以开发基于 Web 技术的网络应用系统。本书介绍的内容就是基于 Java Web 技术开发 Web 应用程序。

1.1　Java Web 概述

Java Web 是 Java EE 的重要核心内容，是基于 Java 技术开发动态 Web 应用程序的一系列技术总称。一个 Java Web 应用系统从结构上包括两个部分：其一，前端服务程序，也称为前端程序、前端或客户端程序。用户通过前端服务程序与系统进行交互。例如，在电子商务系统中，用户通过前端浏览器浏览并选择商品等；其二，后端服务程序，也称为后端程序或后端或服务端。后端服务程序接收从前端服务程序发来的请求信息，对请求信息进行处理，然后将处理结果通过前端服务程序展示给用户。例如，在电子商务系统中，将用户已经选择的商品加入商品购物车，然后对购物车中的商品进行支付结算等。典型的基于 Java Web 技术的网络应用的结构如图 1-1 所示。

图 1-1　Java Web 应用系统结构

1.1.1　前端服务程序

前端服务程序是完成用户与系统进行交互的程序的总称，通过前端服务程序，用户与系统进行信息交互，录入输入数据和显示处理结果等。典型的进行 Web 前端开发技术包括 HTML 页面设计、CSS 布局技术、JavaScript 前端动态页面技术等，除了这些技术外，移动 APP、微信小程序也是前端服务程序开发技术。有关 HTML、CSS、JavaScript、移动 APP 及微信小程序开发相关技术不属于本书的讲解范围,可自行搜索相关资料学习。

1.1.2 后端服务程序

在用户与系统通过前端程序进行信息交互的过程中,前端服务程序需要将用户输入的信息发送给后端服务程序进行处理,再将处理结果返回给前端程序并以适当的形式展示给用户。因此,后端服务程序需要具备如下基本能力:其一,接收从前端服务程序发送的信息;其二,对接收到的用户信息进行处理;其三,将信息处理结果返回给前端程序。为了达成以上三个目标,后端服务程序需要以下技术的支撑。

(1) 应用服务器:也称为 Web 容器(Web container),是执行后端服务程序的机构。后端服务程序都运行在应用服务器中。典型的应用服务器包括 Tomcat 应用服务器、WebLogic 应用服务器、JBoss 应用服务器等。本书使用目前在国内应用较多的 Tomcat 应用服务器。

(2) Servlet 技术:完整的表述应该是 server applet,也就是服务端小程序。它是重要的运行在应用服务器中的程序。Servlet 程序接收前端服务程序发来的请求数据,对数据进行处理,例如,将数据保存到数据库中等,然后将处理结果返回给前端服务程序,进而通过前端服务程序将处理结果展示给用户。

(3) JSP 技术:也就是 Java Server Page(Java 服务端页面)。通过 JSP,可以将经过 Servlet 程序处理的结果以 HTML 页面的形式展示给用户。

(4) Thymeleaf 技术:一种动态页面生成技术。通过 Thymeleaf,可以将经过 Servlet 处理的结果数据以 HTML 页面的形式展示给用户。Thymeleaf 与 JSP 类似,是一种动态页面生成技术。

(5) Session 技术:也就是会话管理技术。鉴于 HTTP 缺乏会话管理技术,Java Web 提供了自己的称为 Session 的会话管理技术来管理多个 HTTP 请求的关联关系。

(6) Java Web 其他相关技术:包括数据库操作技术、文件上传下载技术、异步请求技术、模板技术等。

本书将对以上技术进行详细讲解,以期通过学习本教材,能够开发中等规模的基于 Java Web 的网络应用程序。下面从建立 Java Web 开发环境开始介绍。

1.2 建立 Java Web 开发环境

Java Web 技术体系构建在 Java SE 技术体系基础上。建立 Java Web 开发环境包括三方面的工作:其一,安装 Java SE,也就是 Java JDK;其二,安装 Java 开发环境,本书以目前企业使用较多的 IntelliJ IDEA 为工具;其三,安装 Tomcat 应用服务器。

建立 Java Web 开发环境

建立 Java Web 开发环境说明

1.3 开发第一个 Java Web 程序

现在已经安装了必要的 Java Web 程序开发环境，可以开发 Java Web 应用程序了。作为第一个 Java Web 应用程序例子，现在设计一个简单的登录页面。设计完成的登录页面如图 1-2 所示。

图 1-2　登录页面效果

1.3.1　新建 Java Web 项目

为了设计如图 1-2 所示的登录页面，需要新建一个 Java Web 程序工程。首先启动 IDEA 开发工具，如图 1-3 所示。

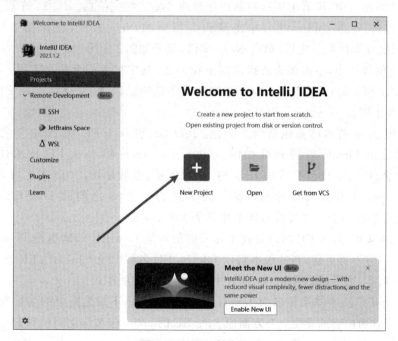

图 1-3　IDEA 启动界面

单击 New Project 按钮，创建一个新的名称为 ch01 的 Java Web 工程，如图 1-4 所示。

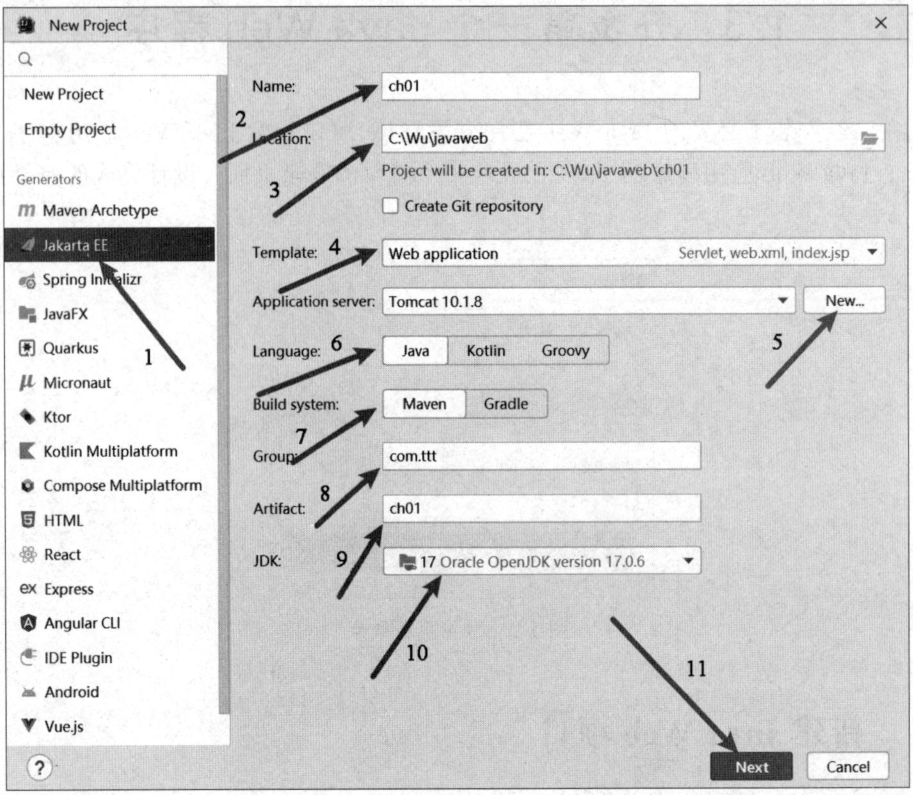

图 1-4　新建名为 ch01 的 Java Web 工程

在图 1-4 的界面中，按箭头所示的顺序选择或者输入相关信息。注意：在箭头 4 指向的下拉列表中选择 Web application；单击箭头 5 指向的 New 按钮，选择在 1.2 节安装的 Tomcat 应用服务器位置。之后，单击 Next 按钮，显示如图 1-5 所示的界面。

在图 1-5 的界面中，单击箭头所指向的下拉列表，其中包含如下几个选择：Java EE 8、Jakarta EE 9.1 和 Jakarta EE 10。为了明确这几个选项的区别，需要对 Java EE 的发展历史有一个初步了解。

如前所述，Java 技术体系包括两个部分：Java SE 和 Java EE，在 2008 年之前，所有的 Java 技术都是由 Orcale 管理和维护的。2008 年，Oracle 将 Java EE 技术体系捐献给 Eclipse 基金会，Oracle 自己只维护 Java SE 技术体系，也就是说，2008 年之后，Java EE 技术体系将由 Eclipse 基金会管理和维护。为了区别于 2008 之前的 Java EE 技术体系，Eclipse 基金会将新的 Java EE 技术体系更名为 Jakarta EE。

截止到 2008 年，Java EE 的最新技术体系的版本是 Java EE 8，这也是 Java EE 的最后一个版本，之后将不再更新，新的版本将以 Jakarta EE 名称出现。Jakarta EE 9.1 是 Jakarta EE 的第一个版本，之后将以 Jakarta 命名并不断推出新的版本。

当然，Java EE 技术体系更名为 Jakarta EE 技术体系，不只是简单的名称变更，还包括以下两个重要方面内容的变更：一方面原 Java EE 的 javax 包名称空间仍归 Oracle，Jakarta EE 不得使用，因此，Jakarta 将原 javax 包名称空间变更为 jakarta；另一方面对于常用的

Tomcat 应用服务器,Tomcat 9.x 是支持 Java EE 技术体系的最后一个版本,从 Tomcat 10.x 开始,Tomcat 应用服务器只支持 Jakarta EE。Java EE、Jakarta EE 与 Tomcat 的版本之间的关系如图 1-6 所示。

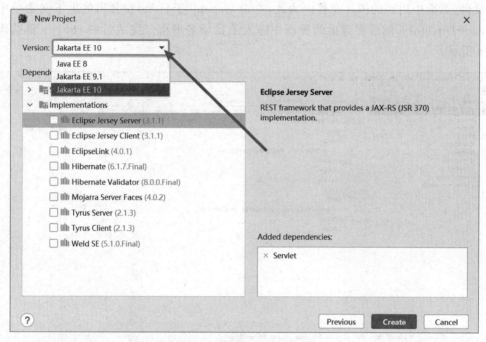

图 1-5　选择 Java Web 的版本

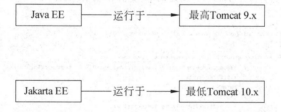

图 1-6　Java EE、Jakarta EE 与 Tomcat 的版本之间的关系

由于 Java EE 不再更新,而 Jakarta EE 才是未来,因此,本书将主要介绍 Jakarta EE。当然,由于本书所介绍的内容是 EE 的核心部分,因此,本书的内容对 Java EE 也是适用的。明确了 Java EE 和 Jakarta EE,现在在图 1-5 的界面箭头所指向的下拉列表中选择 Jakarta EE 10,然后单击 Create 按钮,显示如图 1-7 所示的界面。

现在已经完成了名称为 ch01 的 Java Web 程序工程的创建,下面在 ch01 模块中编写登录界面程序。

1.3.2　编写登录页面代码

编写程序的一个良好习惯是将代码分门别类地放置在工程合适的子目录之中。IDEA 就已经这样做了,IDEA 要求将所有的 Java 代码及其资源放置在"src/main/java"工程目录下;将所有的前端代码,包括 HTML 文件、CSS 文件、JavaScript 文件及用到的其他页面资源文件放置在 webapp 工程目录下。由于在即将编写的登录页面代码中只有前端页面代码，目前还没有涉及 Java 代码，因此，在 ch01 下的 webapp 工程目录下新建如下几个子目录 html、css、js、images，分别放置前端的 HTML 页面文件、页面布局样式文件、JavaScript 代码文件、页面中用到的图片文件。为此，右击 webapp 子目录，在弹出的上下文菜单中选择 New→Directory，然后在弹出的界面中输入子目录名即可。完成后的 ch01 工程结构如图 1-8 所示。

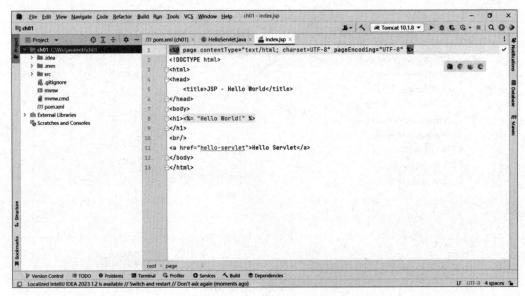

图 1-7　新建的 ch01 Java Web 工程

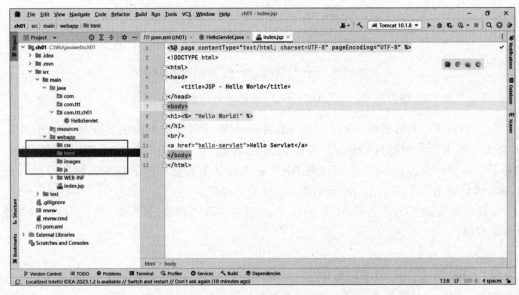

图 1-8　在 webapp 目录下新建子目录

为了使登录页面美观，需要设计 CSS 页面布局文件，在 ch01 工程的 webapp/css 目录下新建一个 style.css 文件。为此，右击 css 子目录，选择 New→Stylesheet，在弹出的界面中

输入文件名 style.css,然后在 style.css 文件中录入如下内容:

```css
input[type=text], input[type=password], select {
    width: 100%;
    padding: 12px 20px;
    margin: 8px 0;
    display: inline-block;
    border: 1px solid #ccc;
    border-radius: 4px;
    box-sizing: border-box;
}
input[type=submit] {
    width: 100%;
    background-color: #4CAF50;
    color: white;
    padding: 14px 20px;
    margin: 8px 0;
    border: none;
    border-radius: 4px;
    cursor: pointer;
}
input[type=submit]:hover {
    background-color: #45a049;
}
div {
    border-radius: 5px;
    background-color: #f2f2f2;
    padding: 20px;
}
```

再创建 HTML 页面代码。将页面代码文件 login.html 放置在 ch01 工程的 webapp/html 目录下。为此,右击 html 子目录,在弹出的菜单中选择 New→HTML File,并在弹出的界面中输入文件名 login.html,然后在 login.html 文件中录入如下代码:

```html
<!DOCTYPE html>
<html lang="en">
<head>
    <meta charset="utf-8">
    <title>Java Web 程序设计</title>
    <link href="../css/style.css" rel="stylesheet" type="text/css" />
</head>
<body>
<h3>请登录</h3>
<div>
    <form action="#">
        <label for="username">用户名</label>
        <input type="text" id="username" name="username" placeholder="用户名...">
        <label for="password">密码</label>
        <input type="password" id="password" name="password" placeholder="密码...">
```

```
            <label for="usertype">用户类型</label>
            <select id="usertype" name="usertype">
                <option value="admin">管理员</option>
                <option value="user">普通用户</option>
            </select>
            <input type="submit" value="提交">
        </form>
    </div>
</body>
</html>
```

完成代码编写后的 ch01 工程的结构如图 1-9 所示。

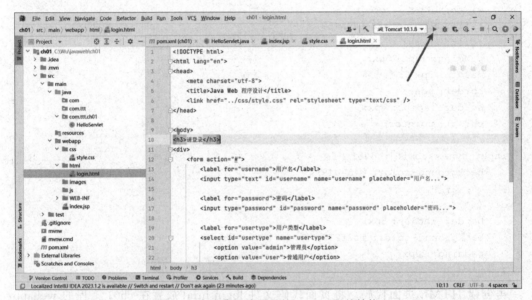

图 1-9　完成代码编写后的 ch01 工程的结构

1.3.3　运行和访问登录页面

为了运行 ch01 程序，在图 1-9 所示的界面中单击箭头所指向的代表"运行"功能的按钮，即可启动程序的运行，如图 1-10 所示。

待 ch01 成功运行后，IDEA 会自动打开浏览器，并显示如图 1-11 所示的界面。

ch01 程序运行后，为什么会自动在浏览器中显示如图 1-11 所示的界面呢？这是 Java Web 程序技术规范的规定：启动一个 Java Web 程序，一旦成功启动，会自动打开 webapp 目录下的名称为 index.html 或者 index.jsp 的称为"主页"的页面。观察图 1-8 或图 1-9 所示的 ch01 工程目录结果，会发现在 webapp 目录下存在名称为 index.jsp 的文件，因此，按照 Java Web 的技术规范要求，将显示 index.jsp 的页面内容，而图 1-11 所显示的内容正是 index.jsp 页面的内容。关于 jsp 页面程序，将在后续内容中介绍。为了显示 login.html 页面内容，在图 1-11 的浏览器地址栏输入 http://localhost:8080/ch01_war_exploded/html/login.html，将显示如图 1-12 所示的页面。

第 1 章　建立 Java Web 开发环境

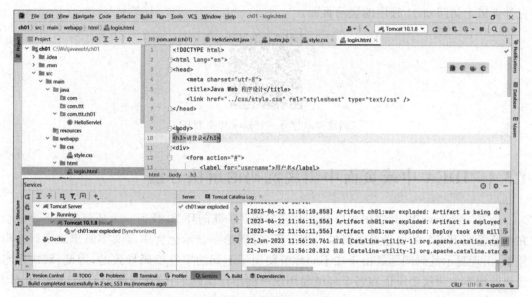

图 1-10　ch01 成功运行后的界面

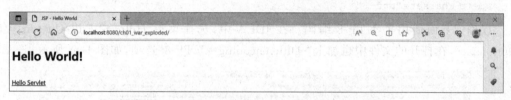

图 1-11　IDEA 自动打开浏览器并显示程序入口页面

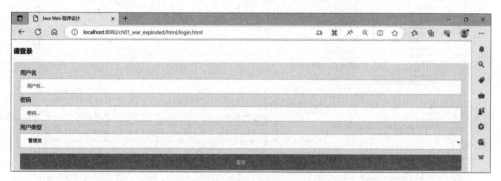

图 1-12　login.html 页面效果

在图 1-12 的页面中，可以在相应的输入框中输入用户名、密码及选择用户类型，然后单击"提交"按钮，此时可以发现浏览器并没有将所输入的用户名、密码及用户类型数据发送到服务器。究其原因是目前这个页面并没有对用户信息进行处理并提交给服务端程序。在后续的章节中会详细介绍如何将信息提交给后端服务程序。

当然，完全可以在另一台计算机的浏览器的地址栏输入 login.html 的页面地址来访问这个登录页面，但是需要将地址 http://localhost:8080/ch01_war_exploded/html/login.html 中的 localhost（因为 localhost 表示运行浏览器的计算机与运行 ch01 程序的计算机是同一台计算机）修改为 ch01 程序运行的计算机的 IP 地址，这时可以发现，完全可以在另一

11

台计算机上访问 ch01 程序的登录页面。

创建和运行第一个 Java Web 程序

1.3.4 IDEA 中或页面中出现乱码的解决方法

开发一个 Java Web 应用程序会涉及许多相互关联的环节，任何一个环节处理不当就会在浏览器或 IDEA 的信息显示窗口出现乱码，特别是在显示包含中文文字时出现乱码。本质上，出现显示乱码问题是由于在不同环节使用了不同的字符编码导致的，因此，解决显示乱码的根本方法就要保证在各个环节使用统一的字符编码，一般建议使用 UTF-8 编码。因此，建议在 HTML 页面文件中，在其 head 标签下的都加上这一句：

```
<meta charset="utf-8">
```

以确保界面使用了 UTF-8 编码。在 IDEA 中，选择菜单 Help→"Edit Custom VM Options..."，在打开的文件中也加上"-Dfile.encoding=UTF-8"选项，如图 1-13 所示。

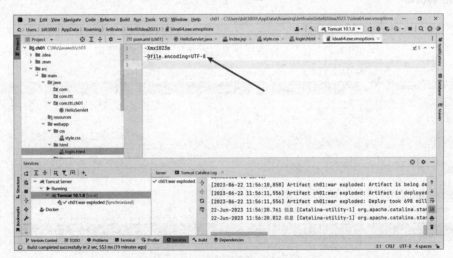

图 1-13 在 IDEA 的配置文件解决控制台乱码问题

提示：在各个环节使用统一的字符编码可以解决信息显示出现时的乱码问题。

1.4 C/S 架构和 B/S 架构

在基于网络的应用中，典型的工作过程是：用户通过计算机向服务器发送请求信息，服务器接收并处理用户发来的信息，并将处理结果返回给用户，然后通过运行在用户计算机上的程序将处理结果显示出来。对于这种信息处理和响应方式,业界称为"客户/服务器"模式,或者称为 Client/Server 模式,简称 C/S 模式。例如,在数据库系统中,用户通过前端程序向数据库服务器发送 SQL 命令,数据库服务器接收并在服务器上执行 SQL 命令,然后将处理结果返回给客户端的程序,并将结果在客户端程序中显示出来。如果客户端计算机上运行的是浏览器,也就是 Browser,例如,Chrome 浏览器或者 Firefox 浏览器等,那么,这种模式就称为"浏览器/服务器"模式,简称为 B/S 模式。从这里也可以看出,B/S 模式是 C/S 模式的一种特殊形式。

1.5 练习:建立 Java Web 开发环境

在自己的计算机上安装 Java Web 程序开发环境,并编写简单的 HTML 页面,测试安装环境的正确性。

第 2 章　HTTP 超文本传输协议

在 B/S 或 C/S 模式的系统中,用户通过运行在客户端计算机上的前端服务程序向后端服务程序发送请求数据,后端服务程序接收并处理请求数据,并将处理结果数据返回给前端服务程序,进而通过前端服务程序将结果显示出来。这里就涉及两台计算机之间的数据通信:前端服务程序是如何向后端服务程序发送请求数据?后端服务程序又是如何接收前端服务程序的请求数据?如何将处理结果数据返回给前端服务程序?在深入学习 Java Web 其他知识之前,需要对前端服务程序与后端服务程序之间的通信有所了解。前端服务程序与后端服务程序之间的通信,简言之,是通过称为 HTTP(hyper text transfer protocol,超文本传输协议)进行信息交互的。

2.1　前端服务程序与后端服务程序之间的通信

在详细介绍 HTTP 之前,首先从直观上观察一下前端服务程序与后端服务程序之间是如何进行通信的,这里的前端服务程序以浏览器为例。后端服务程序以 Web 服务器为例。也就是说,观察一下浏览器是如何向 Web 服务器发送请求信息,又是如何接收从 Web 服务器发送的结果数据。这需要对 HTTP 有初步的了解。

2.1.1　HTTP 通信模型

因特网是全球最大的计算机网络,这个网络已经延伸到了世界的每一个角落。通过因特网,任何联网的计算机都可以进行数据通信。人际之间的沟通需要遵循一定的规则类似,任何两台计算机之间要进行通信也必须遵循一定的规则。这个规则称为"通信协议"。目前最有效、最灵活,也是使用最为广泛的通信协议是 TCP/IP。TCP/IP 不是一个协议,而是一个协议族。TCP/IP 协议族中包含的协议有 ICMP、TCP、IP、UDP、FTP、HTTP 等。每个网络通信协议都定义了任何两台计算机之间进行通信时应该遵循的过程和数据信息格式(也称为消息格式)。HTTP 的通信模型如图 2-1 所示。

在 HTTP 通信过程中,用户通过客户端计算机与服务器端计算机建立网络连接,然后向服务器发送"HTTP 请求"(HTTP request)信息,在 HTTP 请求消息中包括服务器的资源地址和向服务器发送的请求数据;服务器端计算机收到请求消息后,对请求数据进行处理,这些处理包括从数据库中获取数据,或者从服务器指定的地址获取相应资源,然后将这些数据组织成"HTTP 应答"(HTTP response)并发送给客户端计算机;客户端计算机收到服务器的应答消息后,将消息数据以适当的形式在客户端计算机上显示出来。

第 2 章　HTTP 超文本传输协议

图 2-1　HTTP 通信过程

2.1.2　HTTP 通信过程跟踪

在浏览网页的过程中，用户在客户端计算机上运行浏览器程序通过 HTTP 与 Web 服务器之间进行通信。为了对 HTTP 通信过程有一个直观认识，下面使用浏览器提供的工具，并在浏览器上访问服务器上的指定页面资源，对 HTTP 的通信过程进行跟踪。简单起见，以访问百度搜索的首页面为例来跟踪和观察浏览器与服务器之间的 HTTP 通信过程。

HTTP 通信过程跟踪

首先打开浏览器，此处使用 Edge 浏览器。在浏览器地址栏输入百度搜索引擎地址，如图 2-2 所示。此时，显示百度搜索的首页面。

在图 2-2 的界面中，单击箭头所指向的"…"按钮，打开 Edge 浏览器提供的"开发人员工具"功能，如图 2-3 所示。

图 2-2　百度搜索首页面

此时，Edge 浏览器将打开用于辅助开发人员跟踪 HTTP 过程和检查页面属性的"开发人员工具"功能，如图 2-4 所示。

在图 2-4 所示的界面，选择箭头 1 所指向的"网络"选项卡，然后单击箭头 2 所指向的浏览器的"刷新"按钮，将显示如图 2-5 所示的界面。

在图 2-5 所示的界面中，观察由矩形框框住的内容：浏览器为了显示完整的百度搜索首页面，向服务器发送了 61 次请求（注意矩形框底部的 61 这个数字）。单击任何一个请求，可以查看到请求的详细信息。例如，在图 2-5 所示的界面中，单击矩形框中的第一个请求，即可观察到这个请求的详细信息，如图 2-6 所示。

15

图 2-3　打开开发人员工具操作过程

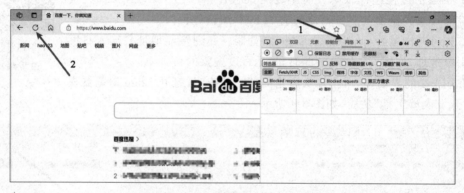

图 2-4　打开后的开发人员工具界面

图 2-5　刷新页面后的结果

图 2-6　一次请求的详细信息

在图 2-6 所示的界面中，观察由矩形框 2 所框住的内容：第 1 行"请求 URL"是说浏览器向服务器 https://www.baidu.com 请求首页面；第 2 行"请求方法"是说浏览器使用称为 GET 的方法向服务器发送数据；第 3 行"状态代码为：200 OK"表示服务器正确地处理了请求，并将结果返回给了浏览器；第 4 行"远程地址"为 14.119.104.189:443，表示百度服务器的 IP 地址和端口。这里简述了本次 HTTP 请求的整体信息。滚动这个小窗口可以观察发送的请求信息和应答信息的详细情况，如图 2-7 和图 2-8 所示。

图 2-7　HTTP 请求的详细信息

在此，不对请求信息和响应信息做详细介绍，只需了解如何跟踪 HTTP 过程即可。在 2.2 节详细介绍了 HTTP 请求消息之后，就可以对协议消息数据格式有更深入理解。

图 2-8　HTTP 应答/响应详细信息

2.2　HTTP 请求消息

为了理解 HTTP 过程跟踪中的数据，需要对 HTTP 请求和 HTTP 应答消息格式有一定的了解。本节介绍 HTTP 请求数据格式。

2.2.1　认识 URL

之所以在浏览器中可以访问网络资源，是因为任何 Web 网络，当然也包括因特网上的任何资源，都有一个唯一的称为 URL 的地址。例如，因特网上的一个页面或者任何网络上的图片都有一个全球唯一的 URL 地址。URL 的全称是 uniform resource locator，中文名称是"统一资源定位器"，有时也称为"全球资源定位器"或者"网址"。在浏览器上浏览一个网页时，在地址栏输入的信息就是页面或资源的 URL 地址。例如，在浏览器地址栏输入百度的首地址，将打开百度的搜索首页面。URL 地址的一般格式如图 2-9 所示。

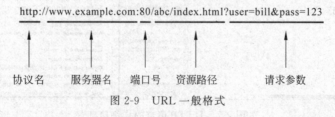

图 2-9　URL 一般格式

在 URL 地址中，协议名 http 可以省略；如果端口号为 80，则端口号可以省略；当请求一个服务器上的主页面时，资源路径也可以省略；对于请求参数部分，不同的 URL 有不同的参数要求。

在一些书籍或文献中,有时会看到称为 URI(uniform resource identifier)的术语,这个术语看起来与 URL 很类似,都是用于标识一个唯一资源。简单理解就是,URL 是 URI 的特殊形式。

2.2.2 HTTP 请求消息格式

了解 URL 之后,下面看看浏览器是如何向服务器发送 HTTP 请求消息的。简单来说,当浏览器要向服务器发送消息时,浏览器会对要发送的消息进行封装。HTTP 定义了浏览器如何封装要发送到服务器的消息的格式。HTTP 请求消息或者称为 HTTP Request 消息的一般格式如图 2-10 所示。

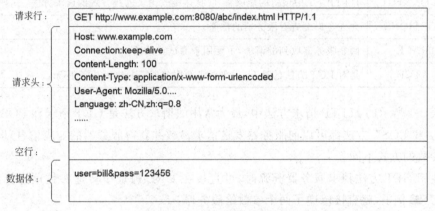

图 2-10　HTTP 请求消息格式

从图 2-10 可以看出,HTTP 请求消息中包括请求行、请求头、空行和数据体四个部分。

(1) 请求行:规定了请求方法(或称为请求方式)、请求的 URL 地址和使用的 HTTP 版本。HTTP 的请求方法包括 GET 请求、POST 请求、DELETE 请求、HEAD 请求、PUT 请求、OPTIONS 请求、TRACE 请求、CONNECT 请求、PATCH 请求。比较常用的请求方法是 GET 请求和 POST 请求;目前常用的 HTTP 版本是 1.1 版本和 2.0 版本。

(2) 请求头:指明了请求的属性相关信息。请求头具有"名:值"对的形式。例如,Host 请求头指明了所请求的目标服务器的名字;Connection 请求头则指明了网络连接的特性;Content-Length 请求头指明了在数据体中数据的长度等。

(3) 空行:用于分隔请求头与数据体数据。

(4) 数据体:有时也称为请求体。当客户端有较多数据需要发送给服务器时,如需要向服务器发送大量的信息时,需要将这些数据放置在"数据体"部分发送。简单地说,就是数据体用于封装客户机浏览器要发送到服务器的数据。

2.2.3 HTTP 请求方法

在 HTTP 的请求消息的请求行中,包含一个重要的参数,这就是"请求方法"。HTTP 定义的请求方法及其功能如表 2-1 所示。

表 2-1 HTTP 请求方法及其含义

序号	请求方法名	功能描述
1	GET	请求指定的网络资源信息,服务器向客户端返回所请求的资源实体
2	HEAD	类似 GET 请求,只不过返回的响应中没有响应体数据,只用于获取响应头信息
3	POST	向服务器指定资源 URL 地址提交数据进行处理,典型的应用是提交表单或者上传文件。数据被包含在请求体中。POST 请求可能会导致新的资源的建立和/或已有资源的修改
4	PUT	从客户端向服务器传送的数据取代指定的文档的内容
5	DELETE	请求服务器删除指定的资源内容
6	CONNECT	HTTP 1.1 中预留的能够将 TCP 连接改为长连接方式的请求方法
7	OPTIONS	允许客户端查看服务器的性能
8	TRACE	回显服务器收到的请求,主要用于测试或诊断
9	PATCH	是对 PUT 方法的补充,用来对已知服务器资源进行局部更新

在表 2-1 所示的 HTTP 请求方法中,最为常用的请求方法是 GET 请求和 POST 请求。GET 方法和 POST 方法都可以向服务器发送请求参数并获得服务器的响应信息,但它们发送请求参数的方式不同。

在使用 GET 方法请求服务器资源时,可直接在 URL 地址中传递参数信息。例如,下面这个 URL 例子,就直接传递了两个参数给服务器:

http://www.example.com:8080/html/abc?username=bill&password=123456

在这个 URL 示例中,直接在"?"后向服务器传递了参数名为 username 且值为 bill 的参数,以及参数名为 password 的值为 123456 的两个参数。GET 方法能够向服务器发送的参数不能超过 1024 字节。一般而言,当浏览器要向服务端程序发送含有包含密码等敏感信息时,建议不要使用 GET 方法,而是使用 POST 方法。

POST 方法则不同。POST 方法发送给服务器的参数不是附加在 URL 地址之后,而是将参数放置在 HTTP 请求的数据体中。因此,使用 POST 方法能够向服务器发送复杂的请求参数,甚至包含文件信息。同时,由于 POST 方法将请求参数放置在 HTTP 的请求数据体中传送给服务器,因此,数据的长度也是不受限制的。

2.2.4 HTTP 常见请求头

在 HTTP 请求消息中,请求头是一个比较复杂的部分。通过请求头,使得客户端可以向服务器发送一些特殊的属性信息。常见的请求头如表 2-2 所示。需要说明的是,除了如表 2-2 所示的常见请求头外,开发人员可以根据需要定义自己的请求头。

在后续适当的时候,会对 HTTP 请求头的含义及作用进行详细介绍。在这个阶段,只需简单了解有这些请求头即可。

表 2-2　HTTP 常见请求头及其含义

序号	请求头名称	含义	举例
1	Accept	向服务器说明客户端可以接受的从服务器返回的响应内容的类型(Content-Types)	Accept：text/plain
2	Accept-Charset	向服务器说明客户端可接受的从服务器返回的字符的字符集类型	Accept-Charset：utf-8
3	Accept-Encoding	向服务器说明客户端可接受的响应内容的编码方式	Accept-Encoding：gzip, deflate
4	Accept-Language	向服务器说明客户端可接受的响应内容语言列表	Accept-Language：zh-CN
5	Connection	向服务器说明客户端优先使用的连接类型	Connection：keep-alive
6	Cookie	向服务器返回之前服务器通过 Set-Cookie 设置的一个 HTTP 的 Cookie	Cookie：$ Version=1；Skin=new
7	Content-Length	请求消息中数据体中以字节个数表示的请求体的长度	Content-Length：1200
8	Content-Type	向服务器说明在 HTTP 请求中请求体的 MIME 类型	Content-Type：application/x-www-form-urlencoded
9	Host	表示服务器的域名以及服务器所监听的端口号。如果所请求的端口是对应的服务的标准端口(80)，则端口号可以省略	Host：www.example.com
10	Referer	表示所访问的前一个 URL 地址，可以认为是之前访问 URL 的链接将客户端带到了当前页面	Referer：http://www.example.com/js/index
11	User-Agent	客户端的身份标识字符串	User-Agent：Mozilla/……（后面太长，省略）

2.3　HTTP 响应消息

当服务器接收到从客户端发送的 HTTP 请求并经过适当处理后,服务器会将请求消息的处理结果返回给客户端。此时,服务器会将要发送给客户端的结果封装成称为"HTTP 响应消息"的格式,然后再发送给客户端。

2.3.1　HTTP 响应消息格式

当服务器需要向客户端发送响应消息(也称为应答消息)时,必须将响应消息按要求封装为 HTTP 响应消息。HTTP 响应消息的一般格式如图 2-11 所示。

从图 2-11 可以看出,HTTP 的响应消息包括状态行、响应头、空行和响应数据体四个部分。

(1) 状态行：简单说明本次请求和应答采用的 HTTP 版本,常见的 HTTP 版本是 HTTP 1.1；本次请求的处理结果称为状态码,用一个编号表示,200 表示正确处理了客户端

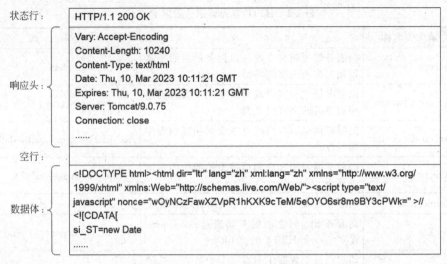

图 2-11　HTTP 响应消息格式

的请求;最后是状态码的简单含义,例如,当结果状态码为 200 时,其含义为 OK。关于更多状态码及其含义在 2.3.2 小节介绍。

(2) 响应头:类似于通过 HTTP 请求头,客户端可以使用 HTTP 请求头告知服务器一些特殊的信息,服务器也可以通过 HTTP 响应头告知客户端一些特殊信息。HTTP 响应头也是"名:值"对的形式。例如,在图 2-11 所示的服务器响应消息中,Content-Length 响应头则告诉客户端,在响应的数据体中的数据长度;Server 响应头则告诉客户端产生这个 HTTP 响应的服务器名称等。关于更多响应头将在 2.3.3 小节介绍。

(3) 空行:用于分隔响应头与数据体数据。

(4) 响应数据体:有时也称为响应体。当服务器有较多数据需要发送给客户端时,此时,需要将这些数据放置在"数据体"部分发送。简单地说,就是数据体用于封装服务器向前端客户端发送响应数据。例如,当客户端浏览器请求某个网络页面时,服务器就将返回的页面 HTML 文件放置在数据体中发送给浏览器。

2.3.2　HTTP 状态码

在 HTTP 的响应消息中,除了 200 这个表示 OK 的状态码外,还有其他几个常见的状态码。这些常见的状态码如表 2-3 所示。

表 2-3　HTTP 状态码及其含义

序号	状态码	状态码英文名称	中文含义
1	200	OK	请求成功。一般用于对 GET 与 POST 请求的应答
2	203	Non-Authoritative Information	非授权信息。请求成功,但返回的元信息不在原始的服务器,而是一个副本
3	301	Moved Permanently	永久移动。请求的资源已被永久地移动到新 URI,返回信息会包括新的 URI,客户端会自动定向到新 URI,今后任何新的请求都应使用新的 URI 代替

续表

序号	状态码	状态码英文名称	中文含义
4	304	Not Modified	未修改。所请求的资源未修改,服务器返回此状态码时,不会返回任何资源。客户端通常会缓存访问过的资源,通过提供一个头信息指出客户端希望只返回在指定日期之后修改的资源
5	400	Bad Request	客户端的请求存在语法错误,服务器无法理解
6	401	Unauthorized	这个请求的 URL 要求用户的身份认证
7	403	Forbidden	服务器理解请求客户端的请求,但是拒绝执行此请求
8	404	Not Found	服务器无法根据客户端的请求找到资源。通过此代码,页面设计人员可设置"您所请求的资源无法找到"的个性页面
9	500	Internal Server Error	服务器内部错误,无法完成请求
10	503	Service Unavailable	由于超载或系统维护,服务器暂时地无法处理客户端的请求。延时的长度可包含在服务器的 Retry-After 头信息中

其中的 404 错误和 500 错误是在开发 Java Web 程序时经常遇到的问题。特别需要注意的是,500 错误出现的原因一般都是因为应用服务器在执行服务端代码时,由于编码者代码的原因而导致服务器发生错误。

2.3.3 HTTP 常见响应头

服务器通过使用 HTTP 响应头来通知客户端关于 HTTP 响应消息的特殊信息。常见的 HTTP 响应头如表 2-4 所示。

表 2-4 常见的 HTTP 响应头

序号	响应头名称	含义	举例
1	Allow	服务器支持哪些请求方法	Allow：GET, HEAD
2	Content-Disposition	对已知 MIME 类型资源的描述,客户端可以根据这个响应头决定是对返回资源的动作。通常用于下载文件时,表示下载文件的类型等	Content-Disposition: attachment; filename="fname.ext"
3	Content-Encoding	数据体中数据所使用的编码类型	Content-Encoding：gzip
4	Content-Language	数据体中数据所使用的语言类型	Content-Language：zh-CN
5	Content-Length	响应体中数据的长度,以字节为单位	Content-Length：1024
6	Content-Type	表示数据体的数据属于什么 MIME 类型,通常需要显式地指定为 text/html	Content-Type：text/html; charset=utf-8
7	Date	当前的 GMT 时间。可以用 setDateHeader 来设置这个头,以避免转换时间格式的麻烦	Date：Tue, 15 Nov 1994 08:12:31 GMT
8	Expires	应该在什么时候认为文档已经过期,从而不再缓存它	Expires：Thu, 01 Dec 1994 16:00:00 GMT
9	Server	服务器名字,由 Web 服务器自己设置	Server：tomcat/9.0.75
10	Set-Cookie	设置和页面关联的 Cookie	Set-Cookie：UserID=itbilu; Max-Age=3600; Version=1

2.3.4 响应体

客户端使用 HTTP 请求消息向服务器请求指定的 URL 资源,服务器经过适当处理后,在 HTTP 响应消息的"响应体"中返回资源给客户端系统,响应体也称为"数据体"。例如,当客户端向服务器请求指定的 HTML 页面时,服务器将客户端所请求的 HTML 资源放置在响应体中返回给客户端;当客户端向服务器请求指定的图片资源时,服务器将相应的图片字节流放置在响应体中并返回给客户端。

2.4 对 HTTP 请求进行深入跟踪剖析

在了解了 HTTP 之后,再次使用浏览器的"开发者工具"跟踪第 1 章所编写的登录页面运行过程。打开 IDEA,并启动 ch01 程序,如图 2-12 所示。

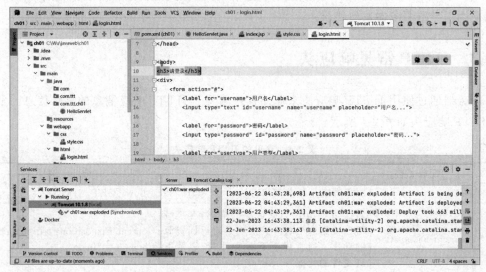

图 2-12 启动登录程序

打开 Edge 浏览器,开启浏览器的"开发者工具",并选择"网络"选项卡。然后在浏览器中输入 URL 地址 http://localhost:8080/ch01_war_exploded/html/login.html,将显示登录页面及浏览器发送给服务器的 HTTP 请求,如图 2-13 所示。

从图 2-13 中可以看出,浏览器向服务器发送了请求 login.html 页面内容的 HTTP 请求,可是,为什么还有箭头 2 所指向的请求 style.css 的 HTTP 请求呢?观察一下 login.html 页面文件就能知道原因。login.html 页面文件如下:

```
<!DOCTYPE html>
<html lang="js">
<head>
    <meta charset="utf-8">
    <title>Java Web 程序设计</title>
```

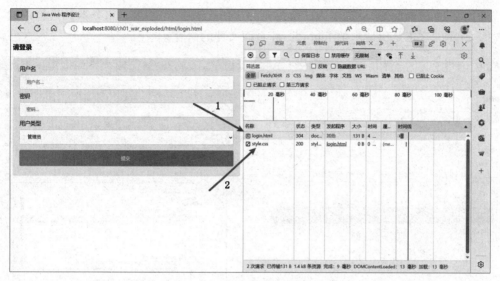

图 2-13　登录页面的 HTTP 请求

```
<link href="../css/style.css" rel="stylesheet" type="text/css" />
</head>
<body>
<h3>请登录</h3>
<div>
...
```

注意其中加黑的代码，这段代码的作用是：浏览器为了正确显示登录页面内容，还需要从服务器上下载针对 login.html 的样式文件。因此，还需要发送 HTTP 请求从服务器请求这个样式文件内容。这就是为什么有两个 HTTP 请求的原因。

现在，单击任何一个 HTTP 请求，例如单击 login.html 这个请求，将显示浏览器请求 login.html 的详细信息，如图 2-14 所示。

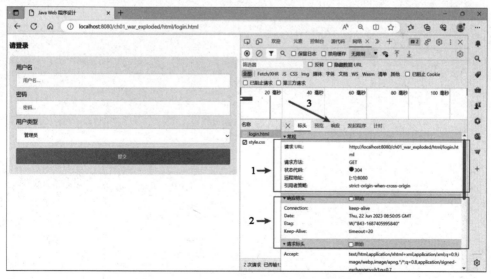

图 2-14　单击 login.html 的详细请求及响应

在图 2-14 的界面中，单击箭头 3 所指向的"响应"选项卡，将显示服务器的应答数据，如图 2-15 所示。这正是所请求的 login.html 页面文件的内容。

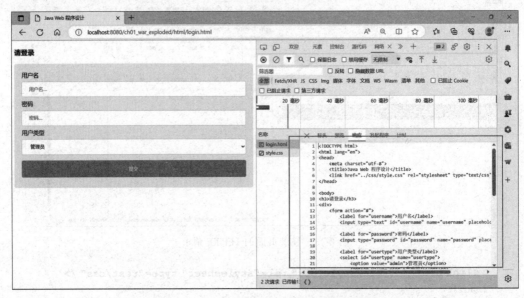

图 2-15　HTTP 应答响应消息的数据体数据

2.5　练习：跟踪浏览器请求和服务器的响应

在第 1 章"建立 Java Web 开发环境"练习的基础上，采用任何你熟悉的浏览器，跟踪浏览器向服务器发送的请求及服务器的响应数据。结合跟踪过程，仔细理解 HTTP 的工作过程。

第 3 章 Servlet 基础

Servlet 是由两个英文单词 Server 与 Applet 构成的复合词，顾名思义，就是服务端程序的意思。这里的服务端程序，也就是前面所说的后端服务程序。既然是后端服务程序，那就要能够接收从前端服务程序发来的请求，进行适当的处理后，将处理结果发送给前端服务程序。本章将对后端服务程序 Servlet 技术进行介绍。

3.1 Servlet 入门

先通过一个简单的例子，直观地看一下 Servlet 后端服务程序包括什么，又是如何工作的。回顾第 1 章，登录页面显示了前端服务程序的登录页面，但是当用户单击"登录"按钮后，程序并没有对登录操作进行处理。本章就从编写一个能够处理登录请求的 Servlet 后端服务程序开始。

3.1.1 完善登录页面

在第 1 章的登录页面中需要将包含用户名、密码和用户类型的参数提交给后端服务程序进行处理，用于验证用户的合法性。先看看登录页面代码 login.html 文件的内容：

```html
<!DOCTYPE html>
<html lang="js">
<head>
    <meta charset="utf-8"/> <title>Java Web 程序设计</title>
    <link href="../css/style.css" rel="stylesheet" type="text/css" />
</head>
<body>
<h3>请登录</h3>
<div>
    <form action="#">
        <label for="username">用户名</label>
        <input type="text" id="username" name="username" placeholder="用户名...">
        <label for="password">密码</label>
        <input type="password" id="password" name="password" placeholder="密码...">
        <label for="usertype">用户类型</label>
        <select id="usertype" name="usertype">
            <option value="admin">管理员</option>
            <option value="user">普通用户</option>
        </select>
```

```
            <input type="submit" value="提交">
        </form>
</div>
</body>
</html>
```

注意：其中的加粗代码<form action="#">是不完整的，需后续修改补充。

为了向后端服务程序提交登录表单数据，需要对这条语句进行修改，需要指明将表单数据发送给服务器上的哪个后端服务程序进行处理，具体采用什么方法向后端服务程序发送信息，以及对发送给后端服务程序的信息采用何种方式进行封装等。为此，修改这句代码如下：

```
<form enctype="application/x-www-form-urlencoded"
        method="post" action="../Login">
```

这条语句的作用是：其一，采用 application/x-www-form-urlencoded 对发送的信息进行封装；其二，采用 post(也就是 POST，大小写一样)方法向后端服务程序发送请求数据；其三，将数据发送到服务器中的称为"../Login"的后端服务程序进行处理。

form 是 HTML 中一个常用的标签，经常被用来向后端服务程序发送请求数据。form 标签具有多个属性，常用属性及其取值如表 3-1 所示。

表 3-1 form 标签常用属性

序号	属性名	常用属性值	含义
1	method	GET	采用 HTTP 的 GET 方法向服务器发送数据
		POST	采用 HTTP 的 POST 方法向服务器发送数据
2	enctype	application/x-www-form-urlencoded	在发送请求数据前对所有特殊字符进行编码
		multipart/form-data	不对字符编码。在需要上传包含文件的表单时必须使用该值
		text/plain	空格转换为"+"，但不对特殊字符编码
3	action	URL 可以是绝对路径，也可以是相对路径	规定当提交表单时向何处发送表单数据
4	target	_blank	在新的浏览器窗口显示服务器返回的结果
		_self	在当前浏览器窗口显示服务器返回的结果

至此，已经完成前端登录页面的修改。下面编写后端服务程序用于接收和处理登录请求。

3.1.2 创建 ch03 工程

按照与第 1 章创建 ch01 工程相似的操作过程创建名为 ch03 的 Java Web 工程。与 ch01 工程类似，在 ch03 工程的 webapp 工程目录下新建 html、css、images、js 子目录，并把 ch01 工程的 webapp 目录下的文件复制到 ch03 的相应目录下。完成 ch03 工程创建后的 IDEA 主界面如图 3-1 所示。

创建了 ch03 工程后，可以编写后端服务程序代码了。

第 3 章　Servlet 基础

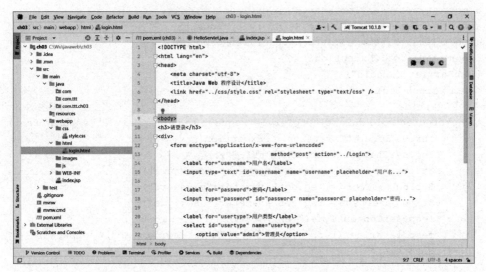

图 3-1　新建的 ch03 工程主界面

3.1.3　创建 Login 后端服务程序

后端服务程序代码是 Java 代码，因此，需要将 Java 代码放置在 ch03 工程的 src→main→java 工程目录下。同时，为了分门别类地管理 Java 代码，在 src→main→java 工程目录下新建一个名为 com.ttt.servlet 的程序包。然后，在这个包下新建一个名为 Login 的 Servlet 程序，此时，ch03 工程的结构如图 3-2 所示。

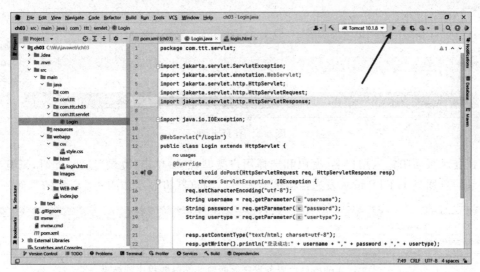

图 3-2　新建 Login 后端服务程序后的 ch03 工程结构

Login.java 代码如下：

```
package com.ttt.servlet;
import jakarta.servlet.ServletException;
import jakarta.servlet.annotation.WebServlet;
```

```java
import jakarta.servlet.http.HttpServlet;
import jakarta.servlet.http.HttpServletRequest;
import jakarta.servlet.http.HttpServletResponse;
import java.io.IOException;
@WebServlet("/Login")
public class Login extends HttpServlet {
    @Override
    protected void doPost(HttpServletRequest req, HttpServletResponse resp)
        throws ServletException, IOException{
        req.setCharacterEncoding("utf-8");
        String username = req.getParameter("username");
        String password = req.getParameter("password");
        String usertype = req.getParameter("usertype");
        resp.setContentType("text/html; charset=utf-8");
        resp.getWriter().println("登录成功:" + username + "," + password + "," +
            usertype);
    }
}
```

现在一切都已经准备就绪，可以运行 ch03 程序了。为此，在图 3-2 所示的界面中，单击箭头所指向的代表运行程序的三角形按钮，将启动 ch03 程序运行。待 ch03 程序启动后，在打开的浏览器中输入如下 URL 地址：http://localhost:8080/ch03_war_exploded/html/login.html，此时将显示登录页面，如图 3-3 所示。

图 3-3　登录页面

在登录页面中输入用户名、密码和选择用户类型，然后单击"登录"按钮，此时，浏览器会将登录信息通过 HTTP 请求发送给服务器，之后，浏览器将显示如图 3-4 所示的界面。

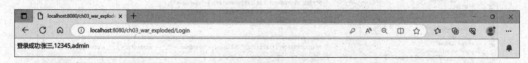

图 3-4　Login 后台服务程序返回给前端服务程序的结果

在浏览器中所显示的信息正是服务端程序发送给浏览器的信息。这正是期望的结果。为什么后端服务程序能够收到前端浏览器发送的登录数据，并且能够将信息返回给浏览器呢？与登录页面进行交互的服务端程序 Login.java 是一个 Servlet 程序。下面详细介绍 Servlet 服务端程序的工作原理。

3.2 Servlet 的具体应用

在详细介绍如何编写后端服务程序之前，需要对 login.html 登录页面与后端服务程序 Login.java 的信息交互过程有一个了解。这里的后端服务程序 Login 是 Servlet 程序，习惯上称为 Login Servlet。

3.2.1 登录页面与后端服务程序 Login 的交互过程

在登录页面中，使用了 form 表单向后端服务程序发送 HTTP 请求。登录页面中 form 表单的关键代码如下：

```html
<form enctype="application/x-www-form-urlencoded" method="post" action="../Login">
    <label for="username">用户名</label>
    <input type="text" id="username" name="username" placeholder="用户名...">
    <label for="password">密码</label>
    <input type="password" id="password" name="password" placeholder="密码...">
    <label for="usertype">用户类型</label>
    <select id="usertype" name="usertype">
        <option value="admin">管理员</option>
        <option value="user">普通用户</option>
    </select>
    <input type="submit" value="提交">
</form>
```

在这段代码中，当用户单击"提交"按钮后，浏览器会将 form 表单的内容采用指定的封装格式 application/x-www-form-urlencoded 进行封装，然后采用 HTTP 请求的 post 方法将请求发送给后端服务程序"../Login"进行处理。为什么在服务程序 Login 的前面会有两个表示父路径的"../"呢？这是因为 login.html 代码部署在 ch03 工程上下文路径"http://localhost:8080/ch03_war_exploded"的 html 子路径下，而 Login 程序则部署在上下文路径的称为根路径的"/"下，因此，在使用相对路径指定目标 URL 时，需要明确目标程序的 URL 地址。

再看看后端服务程序 Login.java 的代码。Login.java 关键代码如下：

```java
@WebServlet("/Login")
public class Login extends HttpServlet{
    @Override
    protected void doPost(HttpServletRequest req, HttpServletResponse resp)
            throws ServletException, IOException{
        req.setCharacterEncoding("utf-8");
        String username = req.getParameter("username");
        String password = req.getParameter("password");
        String usertype = req.getParameter("usertype");
        resp.setContentType("text/html; charset=utf-8");
        resp.getWriter().println("登录成功:" + username + "," + password + "," +
```

```
        usertype);
    }
}
```

先看第一句代码：

```
@WebServlet("/Login")
```

@WebServlet是Java Web规范定义的一个注解（有时也称为标注），这个标注的作用是指定所标注的Servlet程序部署时在Web上下文中的URL路径。括号中的参数"/Login"表示将这个Login Servlet程序部署在Web上下文根路径的Login名称下。也就是说，通过Web上下文路径加上"/Login"，就可以访问这个Servlet程序。而"Web上下文路径"是在部署时指定的。在这个例子中，Web上下文路径为http://localhost:8080/ch03_war_exploded，这是在部署ch03工程时指定的。为了观察一下如何指定一个Java Web程序的部署上下文路径，在IDEA的主界面单击Edit Configuration，如图3-5所示。将打开运行配置界面，如图3-6所示。

图3-5　查看或设置Java Web程序的上下文路径界面

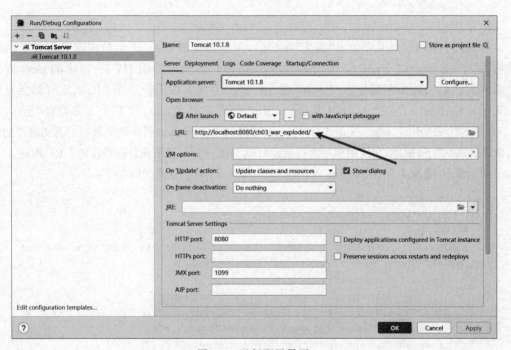

图3-6　运行配置界面

在图 3-6 的界面的箭头所指向的输入框中可以修改程序运行的上下文路径。现阶段建议使用默认值。现在再看看 Login 类的定义代码：

```
public class Login extends HttpServlet{
```

这行代码定义了 Login 类。在 Java Web 中，一般情况下，所有的 Servlet 程序都继承 HttpServlet 这个父类。HttpServlet 类是在 jakarta.servlet-api 这个包中定义的，在 ch03 工程的 pom.xml 文件中导入了这个<dependencies>依赖，如图 3-7 所示。

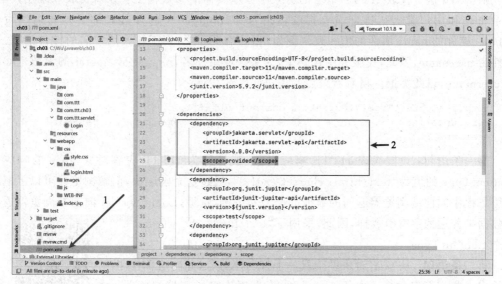

图 3-7　导入 jakarta.servlet-api 包

最后，看看 Login 这个 Servlet 的关键处理代码，也就是其中的 doPost 方法的代码：

```
protected void doPost(HttpServletRequest req, HttpServletResponse resp)
        throws ServletException, IOException{
    req.setCharacterEncoding("utf-8");
    String username = req.getParameter("username");
    String password = req.getParameter("password");
    String usertype = req.getParameter("usertype");
    resp.setContentType("text/html; charset=utf-8");
    resp.getWriter().println("登录成功:" + username + "," + password + "," + usertype);
}
```

这个称为 doPost 的方法具有其固有的含义：当前端登录页面向 Login 这个 Servlet 发送 HTTP 请求时，应用服务器(此处是 Tomcat 10)会将这个请求转交给 Login 程序，因此，Login 会收到这个请求。由于前端页面采用了 Post 方法向 login 发送请求，因此，当 Login 收到这个 HTTP 请求时，会调用 Login 中的 doPost 方法处理这个请求。以此类推，如果前端登录页面采用 Get 方法向 Login 发送 HTTP 请求，则 Login 会调用 doGet 方法来处理这个请求；如果前端页面采用 Put 方法发送 HTTP 请求，则后端服务 Servlet 程序会调用其 doPut 方法处理这个请求，等等。再看看其中的语句：

```
req.setCharacterEncoding("utf-8");
```

这条语句的作用是：当后端服务程序从 req 这个代表 HTTP 请求的对象中获取参数信

息时,如果获取的是字符串数据,则指定字符的编码为 utf-8。由于在登录页面中对页面数据指定了 utf-8 编码,所以,在服务端代码获取参数时也必须指定为 utf-8,否则会出现乱码现象。紧接着的三条语句:

```
String username = req.getParameter("username");
String password = req.getParameter("password");
String usertype = req.getParameter("usertype");
```

则从 req 这个代表 HTTP 请求的对象中获取具体的参数信息:分别获取登录用户名、登录密码和用户类型。特别需要提醒,req.getParameter 中的参数名称必须与前端页面 HTML 控件的 name 属性的值一致。例如,在前端页面中,用于输入用户名的控件的 name 属性为 username,因此,在服务端代码中,为了获取用户名,req.getParameter 的参数也必须为 username,以此类推。再看看最后两条语句:

```
resp.setContentType("text/html; charset=utf-8");
resp.getWriter().println("登录成功:" + username + "," + password + "," + usertype);
```

第一条语句:设置代表 HTTP 响应对象(或者称为 HTTP 应答对象)resp 的响应头 ContentType 的值为"text/html;charset=utf-8",通过这个响应头,前端浏览器可以了解后端服务程序向前端浏览器返回的数据的类型和编码,这里,Login Servlet 向前端浏览器返回的数据是普通的字符串数据,因此,采用了"text/html;charset=utf-8" 这个 ContentType。

跟踪 login.html 页面与 Login Servlet 的交互过程

第二条语句:从 resp 对象中获取一个打印流对象,并向这个打印流输出一些字符串数据。由于 resp 打印流直接连接了前端浏览器,因此,向这个打印流输出的任何信息将直接发送到前端浏览器。

为了进一步观察登录页面 login.html 与后端服务程序 Login 程序的交互过程,打开浏览器的"开发者工具",在登录页面中输入用户名、密码和用户类型,如图 3-8 所示。

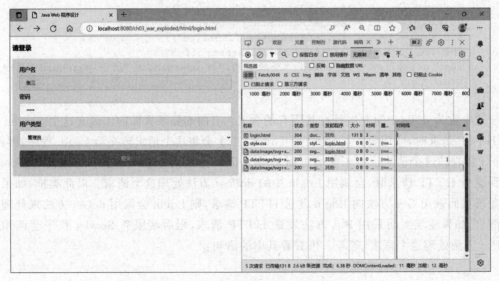

图 3-8 打开浏览器的"开发者工具"

然后单击页面的"提交"按钮，此时，浏览器将登录信息通过 HTTP 请求发送给服务端 Login 程序，Login 程序经过适当处理后，向浏览器返回信息，如图 3-9 所示。

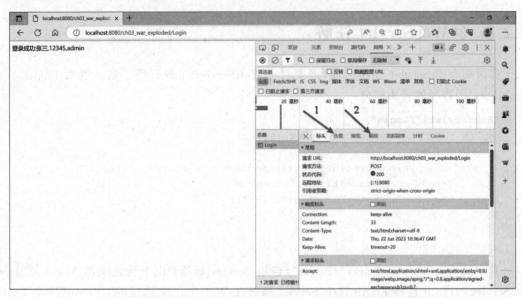

图 3-9　登录页面向 Login 发请求及 Login 返回信息给前端浏览器

在图 3-9 中，单击箭头 1 和箭头 2 所指向的选项卡，将显示浏览器向服务端发送的请求参数信息和从服务端返回的信息，如图 3-10 和图 3-11 所示。

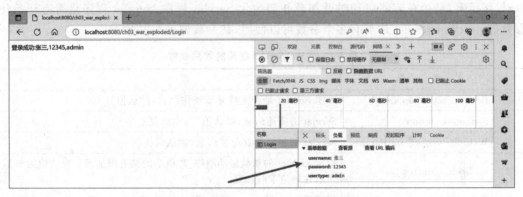

图 3-10　前端浏览器向 Login 发送的请求数据

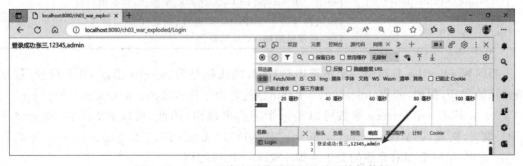

图 3-11　Login 程序向浏览器返回的处理结果数据

至此,已经对前端服务页面与后端服务程序的交互有了比较详细的了解,下面对 Servlet 进行深入介绍。

3.2.2 @WebServlet 注解

在详细介绍 Servlet 程序之前,需要对@WebServlet 注解有所了解。观察 Login.java 这个 Servlet 程序的开头部分:

```
@WebServlet("/Login")
public class Login extends HttpServlet{
    @Override
    protected void doPost(HttpServletRequest req, HttpServletResponse resp)
        throws ServletException, IOException{
            req.setCharacterEncoding("utf-8");
            ...
    }
...
```

其中的代码@WebServlet("/Login")告知 Servlet 容器:以下所定义的 Login 类是一个 Servlet 程序,并且将 Login 这个 Servlet 部署在 Web 上下文路径的"/Login"这个 URL 地址下。

@WebServlet 是 Java Web 定义的一个标准注解,其作用就是使 Servlet 程序可以通过这个注解与 Servlet 容器进行交互,例如,告知 Servlet 容器 Servlet 程序应该部署在哪个 URL 路径下,是否在容器启动时就加载并初始化 Servlet 程序,Servlet 程序是否有初始化参数等。@WebServlet 注解有多个参数可以使用,其参数如表 3-2 所示。

表 3-2 @WebServlet 注解的常用参数

序号	参数	描述
1	Boolean asyncSupported	声明 Servlet 是否支持异步操作模式,默认值为 false
2	String description	Servlet 的描述信息,默认值"",也就是空串
3	WebInitParam[] initParams	Servlet 的初始参数配置信息,默认值为{}
4	int loadOnStartup	指定 Servlet 的加载启动顺序,数值小的被有限加载。默认值为-1,表示在被请求时加载
5	String name	指定 Servlet 的 name 属性,默认值"",也就是空串
6	String[] urlPatterns	指定一组 Servlet 的 URL 路径匹配模式,默认值为{}
7	String[] value	该属性等价于 urlPatterns 属性,两者不能同时指定。如果同时指定,通常是忽略 value 的取值

当@WebServlet 只给出一个字符串参数时,默认就是为 Servlet 指定 URL 路径,这是最常用的方式,例如,@WebServlet("/Login")就等价于@WebServlet(value="/Login")。由于 value 或者 urlPatterns 参数可以是一个字符串数组,因此,可以为 Servlet 指定多个 URL 模式,例如,@WebServlet(value={"/Login","/Login2","/Login/first"})。下面以 Login 为例,给一个相对完整的关于@WebServlet 注解的例子:

```
@WebServlet(value = {"/Login", "/Login/AA", "/Login/bb"},
```

```
    name = "Login Servlet",
    asyncSupported = false,
    loadOnStartup = 10,
    initParams ={
        @WebInitParam(name = "p1", value = "Hello", description = "init 参数 1"),
        @WebInitParam(name = "data", value = "2000", description = "init 参数 2")
    })
    public class Login extends HttpServlet{
        @Override
        protected void doPost(HttpServletRequest req, HttpServletResponse resp)
            throws ServletException, IOException{
                req.setCharacterEncoding("utf-8");
                ...}
    }
```

在这个@WebServlet 注解中，为 Login 指定了三个 URL：/Login、/Login/AA 和/Login/bb，因此，可以使用三个中的任意一个 URL 访问到这个 Servlet 程序，例如，可以在 login.html 代码中，将 form 表单中的 action="../Login"修改为 action="../Login/AA"，仍然可以访问到 Login 这个 Servlet 程序。name 参数为 Servlet 指定一个名字。asyncSupported 参数指定 Servlet 是否支持异步操作模式。loadOnStartup 参数指定 Servlet 在容器启动时被加载。这里主要介绍一下 initParams 参数。

initParams 参数用于为 Servlet 指定初始参数信息，它的值是 WebInitParam 类型的数组，而 WebInitParam 数组元素的值又可以通过@WebInitParam 注解来指定。@WebInitParam 有三个常用参数：name 参数用于指定参数的名称，value 参数用于指定参数的值，description 参数则用于指明参数的含义。例如，在 Login 这个 Servlet 的初始参数配置中指定了两个参数，它们的参数名称分别为 p1 和 data，其值分别为 Hello 和 2000。一旦为 Servlet 指定了初始参数，可以 Servlet 程序中读取相应的参数信息。

3.2.3 Servlet 接口及生命周期

了解了@WebServlet 注解后，可以深入介绍 Servlet 了。本质上，Servlet 是 Java Web 中定义的一个接口。Servlet 接口中定义的方法规范了 Servlet 程序应该具备的基本功能。在 Java Web 文档中，对 Servlet 是这样描述的：Servlet 程序是运行在 Web 服务器中的 Java 程序，它接收从客户端发来的请求并对请求做出响应。Servlet 接口中定义的方法如表 3-3 所示。

表 3-3 Servlet 接口中定义的方法

序号	方 法 名	描 述
1	void init(ServletConfig config)	当应用服务器加载 Servlet 程序时，会调用这个方法初始化 Servlet 程序，其中的 ServletConfig 类型的参数 config 包含了 Servlet 的配置信息，通过这个接口对象可以获得 Servlet 初始化配置等相关信息
2	void service(ServletRequest req, ServletResponse res)	当客户端请求 Servlet 服务时，应用服务器会调用该方法为客户端提供服务

续表

序号	方 法 名	描 述
3	void destroy()	当Servlet被应用服务器移除时,会调用该方法以执行善后处理
4	ServletConfig getServletConfig()	返回表示Servlet配置信息的ServletConfig对象,应用程序可以调用该方法获得Servlet的配置信息
5	String getServletInfo()	返回gaiServlet的基本信息,如作者、版本或版权信息

由于Servlet程序运行在Servlet容器中,因此,Servlet程序的实例化、初始化,为客户端HTTP请求提供服务,以及到最后销毁Servlet实例,都是由Servlet容器控制的。也就是说,Servlet容器管理着Servlet程序的整个存续过程:Servlet接口中定义的init方法、service方法和destroy方法规范了Servlet容器与Servlet程序之间的交互过程,这个过程称为Servlet的生命周期。Servlet的生命周期如图3-12所示。

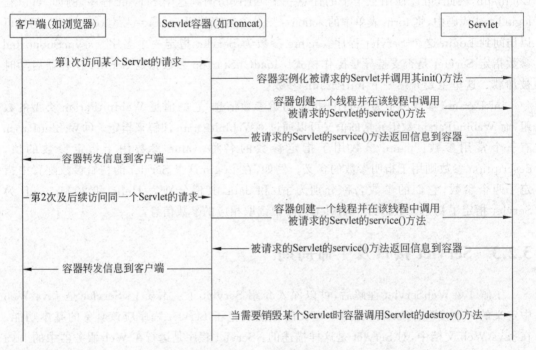

图3-12 Servlet的生命周期

从图3-12可以看出,Servlet的init()方法只在初始化Servlet时被容器执行一次,并且Servlet的destroy()方法也只会在容器需要销毁Servlet时(如容器被下线时)被执行一次,但是Servlet的service()方法会被多次执行。也就是说,对客户端的每次请求,容器都会创建一个新的线程并在新线程中执行Servlet的service()方法。基于Servlet的这种生命周期过程,每个Servlet只会有一个实例,但是,Servlet实例的service()方法会被多个线程同时执行。这种"单例多线程"模式是Servlet的典型特征:在编写Servlet程序时,尤其当存在对共享资源的并发访问时,需要特别留意对资源的并发访问控制。

在Servlet接口的init()方法中,Servlet容器会向Servlet程序传递一个ServletConfig类型的参数,这个参数包含了Servlet初始化配置等相关信息。ServletConfig是一个接口,

Servlet 容器会实现这个接口。ServletConfig 接口定义的方法如表 3-4 所示。

表 3-4　ServletConfig 接口定义的方法

序号	方　法　名	描　　述
1	String getInitParameter(String name)	获取指定 name 的 Servlet 初始配置参数。如果使用@WebServlet 注解为 Servlet 指定了初始参数，则可以通过这个方法获取指定参数的值
2	Enumeration<String> getInitParameterNames()	获取 Servlet 所有的初始配置参数的名称。如果在使用@WebServlet 注解为 Servlet 指定了初始参数，则可以通过这个方法获取所有参数的参数名
3	ServletContext getServletContext()	获取 Servlet 的运行上下文对象
4	String getServletName()	获取 Servlet 的 name 值。如果在使用@WebServlet 注解为 Servlet 指定了 name 参数，则可以通过这个方法获取 name 参数的值

3.2.4　Servlet 接口的实现类 GenericServlet 和 HttpServlet

为了方便编写 Servlet 程序，Java Web 定义了初步实现 Servlet 接口的类，即 GenericServlet。通过继承这个类，可以编写与协议无关的 Servlet 程序。但是，由于目前 Web 前端服务程序与后端服务程序的通信都是通过 HTTP 进行交互的，因此，通过继承 HttpServlet 编写来 Servlet 程序将更为便利。Servlet 接口、GenericServlet 类和 HttpServlet 类之间的关系如图 3-13 所示。

图 3-13　Servlet 接口及其实现类

因为 GenericServlet 是 Servlet 接口的实现类，因此，它实现了 Servlet 接口的所有方法，同时，还增加了一些便利性方法。GenericServlet 类的方法如表 3-5 所示。

表 3-5　GenericServlet 类的方法

序号	方　法　名	描　　述
1	Servlet 接口定义的所有方法	GenericServlet 类实现了 Servlet 接口定义的所有方法
2	String getInitParameter(Stringname)	获取指定 name 的 Servlet 初始配置参数。如果再使用@WebServlet 注解为 Servlet 制定了初始参数，则可以通过这个方法获取指定参数名的值
3	Enumeration<String> getInitParameterNames()	获取 Servlet 所有的初始配置参数的名称。如果在使用@WebServlet 注解为 Servlet 制定了初始参数，则可以通过这个方法获取所有参数的参数名
4	ServletContext getServletContext()	获取 Servlet 的运行上下文对象
5	void log(String msg)	将日志信息写入 Servlet 容器的日志文件中

HttpServlet 是 GenericServlet 类的子类，它针对 HTTP 对 GenericServlet 类进一步优化，使得编写基于 HTTP 的 Servlet 更为便利。HttpServlet 类的常用方法如表 3-6 所示。

表 3-6 HttpServlet 类的常用方法

序号	方法名	描述
1	GenericServlet 类的所有方法	HttpServlet 类继承了 GenericServlet 类的所有方法
2	void service(ServletRequest req, ServletResponse res)	对客户端的 HTTP 请求进行分析，根据请求方式将请求分派到相应的 do×××()方法中进行处理
3	protected void doDelete(HttpServletRequest req, HttpServletResponse resp)	被 service()方法调用，用于对 HTTP 的 DELETE 请求进行处理
4	protected void doGet(HttpServletRequest req, HttpServletResponse resp)	被 service()方法调用，用于对 HTTP 的 GET 请求进行处理
5	protected void doHead(HttpServletRequest req, HttpServletResponse resp)	被 service()方法调用，用于对 HTTP 的 HEAD 请求进行处理
6	protected void doOptions(HttpServletRequest req, HttpServletResponse resp)	被 service()方法调用，用于对 HTTP 的 OPTIONS 请求进行处理
7	protected void doPost(HttpServletRequest req, HttpServletResponse resp)	被 service()方法调用，用于对 HTTP 的 POST 请求进行处理
8	protected void doPut(HttpServletRequest req, HttpServletResponse resp)	被 service()方法调用，用于对 HTTP 的 PUT 请求进行处理
9	protected void doTrace(HttpServletRequest req, HttpServletResponse resp)	被 service()方法调用，用于对 HTTP 的 TRACE 请求进行处理

在编写 Servlet 程序时，最为常用的方式是通过继承 HttpServlet 类，并在子类中重写 do×××()方法。一般而言，HTTP 的 GET 请求和 POST 请求是最为常用的 HTTP 请求，因此，在编写 Servlet 程序时，经常会重写 doGet()方法和 doPost()方法以对客户端的 HTTP 请求进行处理。当然，根据需要，可能还需要重写 init()方法。

至此，已经介绍了 Servlet 的最基础的内容，下面编写一个更为完整的 Login Servlet 程序来处理来自客户端的请求。

3.3 案例：更为完整的 Login Servlet 程序

为了更好地理解和掌握以上内容，现在编写更为完整的 Login 程序，它接收和处理来自前端服务程序请求的同时，通过 Servlet 容器（这里是 Tomcat 10）记录程序的运行日志。通过这个例子，可以了解到如何将运行日志写入 Tomcat 10 的日志文件。

3.3.1 案例目标

所编写的 Login Servlet 程序能满足以下要求：①Login 不仅能够处理客户端的 POST

方法请求,也要能处理客户端的 GET 方法请求;②Login 能够处理可配置的合法用户名和密码等信息;③Login 将客户端的请求信息写入日志文件中。

3.3.2 案例分析

由于新的 Login 程序需要能够处理 HTTP 的 GET 请求和 POST 请求,因此,需要重写 doGet()方法和 doPost()方法;同时,因为新的 Login 程序需要能够处理可配置的合法用户名和密码等信息,因此,可以通过@WebServlet 注解的 initParams 参数来配置合法的用户名和密码;最后,由于需要将客户端的请求信息写入日志文件,可以使用 GenericServlet 类的 log()方法将信息写入日志文件中。

3.3.3 案例实施

为了区别于前面的 Login 程序,在这里把新的 Login 程序命名为 Login2。Login2 程序的代码如下:

```
package com.ttt.servlet;
import jakarta.servlet.ServletConfig;
import jakarta.servlet.ServletException;
import jakarta.servlet.annotation.WebInitParam;
import jakarta.servlet.annotation.WebServlet;
import jakarta.servlet.http.HttpServlet;
import jakarta.servlet.http.HttpServletRequest;
import jakarta.servlet.http.HttpServletResponse;
import java.io.IOException;
import java.util.ArrayList;
import java.util.Collections;
import java.util.List;
@WebServlet(value = "/Login2", initParams = {
    @WebInitParam(name = "users", value = "张三,Bill", description = "用逗号分隔的多个用户名"),
    @WebInitParam(name = "passwords", value = "12345,hello"),
    @WebInitParam(name = "usertypes", value = "admin,user")}
)
public class Login2 extends HttpServlet{
    private List<String> users, passwords, usertypes;
    private int count;
    @Override
    public void init(ServletConfig config) throws ServletException{
        super.init(config);
        log("调用了 init 方法");
        users = new ArrayList<>();
        passwords = new ArrayList<>();
        usertypes = new ArrayList<>();
        String s1 = getInitParameter("users");
        String[] ss = s1.split(",");
        Collections.addAll(users, ss);
        s1 = getInitParameter("passwords");
        ss = s1.split(",");
```

```java
            Collections.addAll(passwords, ss);
            s1 = getInitParameter("usertypes");
            ss = s1.split(",");
            Collections.addAll(usertypes, ss);
            count = users.size();
        }
        @Override
        protected void doGet(HttpServletRequest req, HttpServletResponse resp)
                                        throws ServletException, IOException{
            req.setCharacterEncoding("utf-8");
            String username = req.getParameter("username");
            String password = req.getParameter("password");
            String usertype = req.getParameter("usertype");
            for(int i=0; i<count; i++){
                if(username.equals(users.get(i)) &&
                    password.equals(passwords.get(i)) &&
                    usertype.equals(usertypes.get(i))) {
                    resp.setContentType("text/html; charset=utf-8");
                    String mess = "登录成功:" + username + "," + password + "," +
                        usertype;
                    resp.getWriter().println(mess);
                    log(mess);
                    return;
                }
            }
            resp.setContentType("text/html; charset=utf-8");
            String mess = "登录失败:" + username + "," + password + "," + usertype;
            resp.getWriter().println(mess);
            log(mess);
        }
        @Override
        protected void doPost(HttpServletRequest req, HttpServletResponse resp)
                throws ServletException, IOException{
            doGet(req, resp);
        }
}
```

在 Login2 程序中，首先使用@WebServlet 注解配置了该服务端程序的相关信息，然后在它的初始方法 init()中，在写入一条信息到日志文件后，从配置中获取相关信息，包括合法用户名、密码和用户类型信息，并将它们保存到各自的 List 对象中。

由于在 Login2 中需要能够处理客户端 HTTP 的 GET 请求和 POST 请求，因此，需要重写 doGet()方法和 doPost()方法。但是，在这个例子中，由于 doGet()方法和 doPost()方法对请求的处理过程是一样的，因此，在 doPost()方法中直接调用了 doGet()方法。在处理了客户端请求后，将处理结果信息发送给客户端的同时，也写入了日志文件中。

现在编写客户端页面代码。类似地，为了区别于前面的 login.html 代码，把这个新的页面程序命名为 login2.html。login2.html 的代码如下：

```html
<!DOCTYPE html>
<html lang="js">
<head>
    <meta charset="utf-8">
    <title>Java Web 程序设计</title>
```

```html
        <link href="../css/style.css" rel="stylesheet" type="text/css" />
</head>
<body>
<h3>请登录</h3>
<div>
    <form enctype="application/x-www-form-urlencoded"
        method="GET" action="../Login2">
        <label for="username">用户名</label>
        <input type="text" id="username" name="username" placeholder="用户名...">
        <label for="password">密码</label>
        <input type="password" id="password" name="password" placeholder="密码...">
        <label for="usertype">用户类型</label>
        <select id="usertype" name="usertype">
            <option value="admin">管理员</option>
            <option value="user">普通用户</option>
        </select>
        <input type="submit" value="提交">
    </form>
</div>
</body>
</html>
```

提示：代码中粗体字语句应引起重视。

该语句采用 GET 方法向 Login2 发送 HTTP 请求。现在启动 ch03 程序运行，在浏览器中输入 login2.html，然后在登录界面中输入用户名、密码等信息并单击"提交"按钮，运行结果如图 3-14 所示。

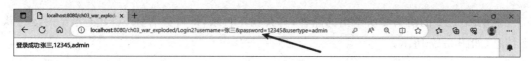

图 3-14　使用 GET 方法发送 HTTP 请求

由于采用了 GET 方法发送请求，因此，用户名、密码和用户类型参数都直接在浏览器的地址栏显示出来了，这是非常不安全的。现在，将黑体的 form 表单修改为如下语句：

```html
<form enctype="application/x-www-form-urlencoded"
        method="POST" action="../Login2">
```

也就是采用 POST 方法发送 HTTP 请求。再次运行程序，在登录界面中输入用户名、密码等信息并单击"提交"按钮，运行结果如图 3-15 所示。

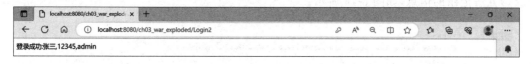

图 3-15　使用 POST 方法发送 HTTP 请求

由于采用 POST 方法发送请求，HTTP 请求中的信息是通过 HTTP 的数据体封装后发送到服务端的，因此，在浏览器地址栏不再显示请求的参数信息。

Login2 程序代码中，在其 init() 方法和 doGet() 方法中都通过 log() 方法将信息写入日志文件中。那么，日志文件具体在哪里呢？要知道日志文件的位置，需要了解 IDEA 启动

Servlet 容器的运行机理。

在 IDEA 中要启动一个 Java Web 程序运行,正如在第 1 章所介绍的,需要为程序配置 Servlet 容器(本书使用的是 Tomcat)相关信息。配置好 Servlet 容器后,IDEA 在启动程序运行时,会为程序创建一个独立的运行环境,这个运行环境在程序运行时,会显示在 IDEA 的 service 选项卡中,如图 3-16 所示。

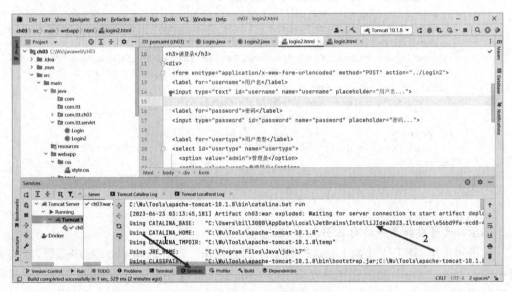

图 3-16 Servlet 容器启动信息

根据图 3-16 的箭头 2 所给出的路径,打开指定目录,即可查看到 Servlet 容器的运行状态信息,如图 3-17 所示。

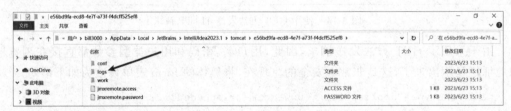

图 3-17 Servlet 运行信息目录内容

打开 logs 子目录,在这个目录下的文件记录了 Servlet 运行相关信息和应用程序通过 log()方法记录的信息,如图 3-18 所示。

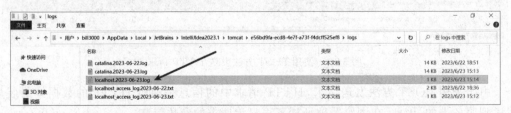

图 3-18 Servlet 日志文件

打开 localhost.2023-06-23.log 文件,即可看到 Login2 程序记录的日志信息。

3.4 HttpServletRequest 对象及其使用

当客户端通过 HTTP 向服务端程序发送请求时,Servlet 容器会将客户端的请求封装成 HttpServletRequest 对象并作为 do×××() 方法的参数传递给后端服务程序。后端服务程序通过 HttpServletRequest 对象可执行一系列操作,例如,获取客户端的请求参数及获取请求头相关信息等。HttpServletRequest 接口是 ServletRequest 的子接口。HttpServletRequest 接口常用的方法如表 3-7 所示。

表 3-7 HttpServletRequest 接口的常用方法

序号	方法名	描述
1	void setCharacterEncoding(String env)	设置请求体中的字符数据的编码
2	String getParameter(String name)	获取请求中指定参数 name 对应的值
3	Enumeration<String> getParameterNames()	返回请求中的所有参数的名称
4	String[] getParameterValues(String name)	如果请求中参数具有多个值,则返回 name 参数对应的所有值
5	String getHeader(String name)	返回 HTTP 请求中指定 name 的请求头信息
6	Enumeration<String> getHeaderNames()	返回 HTTP 请求中所有请求头的名称
7	Enumeration<String> getHeaders(String name)	如果一个请求头具有多个值,则返回这个请求头的所有值信息
8	Object getAttribute(String name)	返回请求对象中指定 name 属性所对应的值,如果不存在 name 属性,则返回 null
9	void setAttribute(String name, Object o)	设置一个 name 属性的值为给定参数对象 o
10	Enumeration<String> getAttributeNames()	返回请求对象中所有属性的 name 值
11	String getCharacterEncoding()	返回请求体中的数据的字符编码名称
12	int getContentLength()	返回请求体中以字节为单位的数据长度
13	String getContentType()	返回请求体中数据的 MIME 类型
14	ServletInputStream getInputStream()	返回一个以 HTTP 请求体中的数据为基础的数据流
15	BufferedReader getReader()	对请求体中数据为字符数据,则返回一个 Reader 对象
16	String getRemoteAddr()	返回客户端的 IP 地址
17	RequestDispatcher getRequestDispatcher(String path)	以指定的 URL 路径作为参数,返回一个分发对象,这个对象可用户转发请求到通过 path 指定的 URL
18	ServletContext getServletContext()	返回程序所运行的 Servlet 容器上下文对象
19	Cookie[] getCookies()	返回 HTTP 请求的所有 Cookies
20	HttpSession getSession(boolean create)	返回与当前 HTTP 请求关联的会话对象,如果会话存在的话,否则如参数为 true,则创建一个新会话

3.4.1 获取请求参数

当前端服务程序向后端服务程序发送 HTTP 请求时,经常会携带一些请求参数。就像前面的登录页面程序向后端服务程序发送请求一样:登录页面向后端服务程序发送了用户名、密码和用户类型参数。后端服务程序则从 HttpServletRequest 对象中使用如下代码获取请求参数信息:

```
req.setCharacterEncoding("utf-8");
String username = req.getParameter("username");
String password = req.getParameter("password");
String usertype = req.getParameter("usertype");
```

为了能够正确地从 HttpServletRequest 对象中获得请求参数,需要使用语句:

```
req.setCharacterEncoding("utf-8");
```

首先设置请求参数的字符编码,这里所设置的编码必须与前端服务程序设置的编码保持一致:在页面中使用如下语句指定了页面的编码:

```
<head>
    <meta charset="utf-8">
    ...
</head>
```

一旦正确地指定了编码,可以使用 HttpServletRequest 对象的 getParameter(String name)方法获取指定参数的值。

当然,也可以使用 HttpServletRequest 对象的 getParameterNames()方法得到所有请求参数的名称。为此,新建 Login3 Servlet 程序,代码如下:

```
package com.ttt.servlet;
import jakarta.servlet.ServletException;
import jakarta.servlet.annotation.WebServlet;
import jakarta.servlet.http.HttpServlet;
import jakarta.servlet.http.HttpServletRequest;
import jakarta.servlet.http.HttpServletResponse;
import java.io.IOException;
import java.util.Enumeration;
@WebServlet(value = "/Login3")
public class Login3 extends HttpServlet{
    @Override
    protected void doGet(HttpServletRequest req, HttpServletResponse resp)
        throws ServletException, IOException{
        req.setCharacterEncoding("utf-8");
        Enumeration<String> names = req.getParameterNames();
        StringBuilder info = new StringBuilder();
        while(names.hasMoreElements()){
            String name = names.nextElement();
            info.append(name).append(":").append(req.getParameter(name));
            info.append("<br>");
        }
```

```
        resp.setContentType("text/html; charset=utf-8");
        resp.getWriter().println(info);
    }
    @Override
    protected void doPost(HttpServletRequest req, HttpServletResponse resp)
            throws ServletException, IOException{
        doGet(req, resp);
    }
}
```

现在,新建页面程序 login3.html,这个页面只在 login2.html 的基础上,修改 form 表单 action 属性的值为新建的 Login3 程序:

```
<formenctype="application/x-www-form-urlencoded"
          method="POST" action="../Login3">
```

运行这个程序,并在浏览器中打开 login3.html 页面,如图 3-19 所示。

图 3-19 login3.html 登录页面

在图 3-19 中,单击"提交"按钮,将显示如图 3-20 所示的页面。

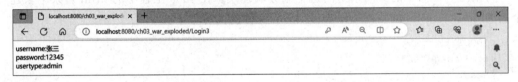

图 3-20 使用 getParameterNames 获取所有参数名

从图 3-20 中可以看出,Login3 服务端程序正确获取了所有请求参数名及其对应的参数值。

3.4.2 获取 HTTP 请求头信息

通过第 2 章介绍 HTTP,可以了解到:一个完整的 HTTP 请求除了包含请求的 URL 地址和请求参数外,还包括 HTTP 请求头。HTTP 请求头是由一系列"名:值"对组成。HttpServletRequest 对象提供了类似 getHeader(String name)方法从 HTTP 请求中获取请求头信息。下面举一个例子,介绍如何获取 HTTP 请求头信息。为此,新建名称为 Headers 的 Servlet 程序,Headers 程序代码如下:

```java
package com.ttt.servlet;
import jakarta.servlet.ServletException;
import jakarta.servlet.annotation.WebServlet;
import jakarta.servlet.http.HttpServlet;
import jakarta.servlet.http.HttpServletRequest;
import jakarta.servlet.http.HttpServletResponse;
import java.io.IOException;
import java.util.Enumeration;
@WebServlet("/Headers")
public class Headers extends HttpServlet{
    @Override
    protected void doGet(HttpServletRequest req, HttpServletResponse resp)
      throws ServletException, IOException{
        String ua = req.getHeader("user-agent");
        resp.setContentType("text/html; charset=utf-8");
        resp.getWriter().println(ua+"<br><br>");
        Enumeration<String> names = req.getHeaderNames();
        StringBuilder info = new StringBuilder();
        while(names.hasMoreElements()){
            String name = names.nextElement();
            info.append(name).append(":").append(req.getHeader(name));
            info.append("<br>");
        }
        resp.getWriter().println(info);
    }
}
```

在这个程序中，先通过语句"String ua = req.getHeader("user-agent");"直接获得uuser-agent请求头信息，然后在页面中显示出来。然后，通过语句"Enumeration<String> names = req.getHeaderNames();"获得所有请求头的名称，并进一步通过循环语句得到各个请求头的值并显示在页面中。

再编写一个简单的页面来访问这个服务端程序。在ch03工程下的html目录下新建一个名为headers.html的页面。headers.html文件的内容如下：

```html
<!DOCTYPE html>
<html lang="en">
<head>
    <meta charset="UTF-8">
    <title>显示所有请求头信息</title>
</head>
<body>
<a href="../Headers">显示所有请求头信息</a>
</body>
</html>
```

运行这个程序，并在浏览器中打开headers.html页面，如图3-21所示。

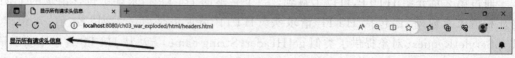

图3-21 显示请求头信息页面

单击"显示所有请求头信息"链接,即可在新的页面中显示所有的请求,如图3-22所示。

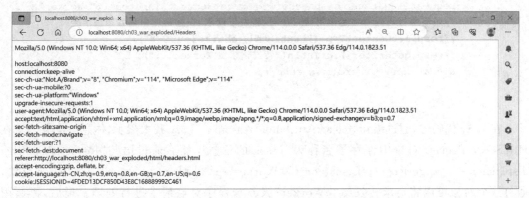

图3-22 获取HTTP的所有请求头信息

也可以通过浏览器直接访问这个服务端程序,访问方式为:ch03工程部署的上下文路径＋"/Headers"。结果与图3-22一致。

从图3-22中可以看出,程序能够正确获取HTTP的所有请求头信息。在编程实践中,服务端可以要求客户端通过HTTP的请求头携带一些特殊的信息,如令牌信息等,以确保信息的安全性。服务端代码通过检查是否存在指定的请求头信息以验证用户访问的合法性。

3.4.3 转发请求和页面包含

在编程实践中,可能存在这样一种情况:客户端请求服务端的某个程序提供服务,可是,所请求的服务端程序目前不能为这个请求提供服务。为了处理这种情况,一种可行的方法是将这个请求转发给其他能够提供服务的程序进行处理。这种方式称为"转发请求",有时也称为"请求转发"。在转发请求时,可以携带一些属性信息到提供服务的程序。先看一个简单的例子。

在这个例子中,名为Lazy的服务端程序接收到客户端的请求后,在HttpServletRequest对象中保存一些信息,然后将这个请求转发给名为Worker的服务端程序进行处理。Lazy程序代码如下:

```
package com.ttt.servlet;
import jakarta.servlet.RequestDispatcher;
import jakarta.servlet.ServletException;
import jakarta.servlet.annotation.WebServlet;
import jakarta.servlet.http.HttpServlet;
import jakarta.servlet.http.HttpServletRequest;
import jakarta.servlet.http.HttpServletResponse;
import java.io.IOException;
@WebServlet("/Lazy")
public class Lazy extends HttpServlet{
    @Override
    protected void doGet(HttpServletRequest req, HttpServletResponse resp)
        throws ServletException, IOException{
```

```java
        req.setAttribute("reason", "Lazy我太累了,休息一会儿");
        RequestDispatcher dispatcher = req.getRequestDispatcher("/Worker");
        //如下两行代码不会起作用,因为一旦转发了请求,在输出流中内容将被清空
        //resp.setContentType("text/html; charset=utf-8");
        //resp.getWriter().println("转发请求给Worker...");
        dispatcher.forward(req, resp);
    }
}
```

在Lazy代码中,通过语句"req.setAttribute("reason","Lazy我太累了,休息一会儿");"在HttpServletRequest对象中保存了名称为reason的信息。然后通过语句"RequestDispatcher dispatcher = req.getRequestDispatcher("/Worker");"得到一个RequestDispatcher对象,通过这个对象可以将请求转发给"/Worker"服务端程序对该请求进行处理。最后,再调用RequestDispatcher对象的forward()方法将请求转发给"/Worker"进行处理。

提示:其中的三条粗体字语句应引起重视。

当把请求转发给一个新的服务端程序处理时,之前向输出流中写入的任何信息将被清空,因此,在Lazy中向客户端写入的信息将不会发送给客户端程序。

现在编写Worker Servlet代码,Worker程序代码如下:

```java
package com.ttt.servlet;
import jakarta.servlet.ServletException;
import jakarta.servlet.annotation.WebServlet;
import jakarta.servlet.http.HttpServlet;
import jakarta.servlet.http.HttpServletRequest;
import jakarta.servlet.http.HttpServletResponse;
import java.io.IOException;
@WebServlet("/Worker")
public class Worker extends HttpServlet{
    @Override
    protected void doGet(HttpServletRequest req, HttpServletResponse resp)
      throws ServletException, IOException{
        String reason = (String)req.getAttribute("reason");
        String info = "";
        if (reason == null){
            info = "客户端直接请求我服务";
        }
        else{
            info = "我接替Lazy工作,因为: " + reason;
        }
        resp.setContentType("text/html; charset=utf-8");
        resp.getWriter().println(info);
    }
}
```

在Worker服务代码中,首先通过语句"String reason =(String)req.getAttribute("reason");"从HttpServletRequest对象中获取名称为reason的属性值。如果这个值存在,则表示这个请求是从Lazy服务程序转发过来的,否则表示这个请求是直接从前端服务程序发来的。运行这个程序,在浏览器中请求Lazy服务,得到如图3-23所示的结果。

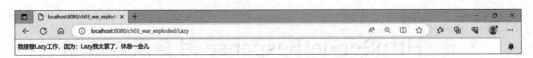

图 3-23　转发 HTTP 请求运行结果

当然，也可以直接从浏览器中请求 Worker 提供服务，为此，在浏览器地址栏直接输入 Worker 的 URL 地址，运行结果如图 3-24 所示。

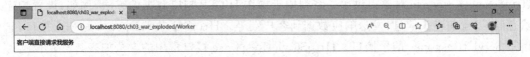

图 3-24　直接请求 Worker 提供服务

服务端 Servlet 程序除了可以将 HTTP 请求转发给另一个 Servlet 程序进行处理外，还可以将另一个 Servlet 的处理结果包含在自己发送给客户端程序的结果中。这种方式称为"页面包含"。为了理解页面包含的含义，新建一个后端服务程序 Lazy1，Lazy1 程序的代码如下：

```
package com.ttt.servlet;
import jakarta.servlet.RequestDispatcher;
import jakarta.servlet.ServletException;
import jakarta.servlet.annotation.WebServlet;
import jakarta.servlet.http.HttpServlet;
import jakarta.servlet.http.HttpServletRequest;
import jakarta.servlet.http.HttpServletResponse;
import java.io.IOException;
@WebServlet("/Lazy1")
public class Lazy1 extends HttpServlet{
    @Override
    protected void doGet(HttpServletRequest req, HttpServletResponse resp)
        throws ServletException, IOException{
        req.setAttribute("reason", "Lazy我太累了,休息一会儿");
        RequestDispatcher dispatcher = req.getRequestDispatcher("/Worker");
        //由于采用页面包含,因此,可以将被包含的服务端程序的输出包含到这个程序中
        resp.setContentType("text/html; charset=utf-8");
        resp.getWriter().println("包含 Worker 的输出信息开始.<br>");
        dispatcher.include(req, resp);
        resp.getWriter().println("<br>包含 Worker 的输出信息结束");
    }
}
```

在 Lazy1 代码中，粗体字语句使用页面包含将 Worker 的输出纳入自己的页面中。此时，不仅可以在 Lazy1 中输出信息到客户端，也可以将被包含的后端服务程序中输出的信息发送到客户端。运行这个程序，访问 Lazy1 服务端程序，运行结果如图 3-25 所示。

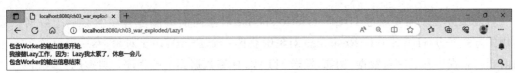

图 3-25　页面包含运行结果

3.5 HttpServletResponse 对象及其应用

HttpServletResponse 对象代表的是后端服务程序向前端服务程序发送的 HTTP 响应；后端服务程序正是通过这个对象向前端服务程序发送对请求的处理结果。HttpServletResponse 对象提供了一系列用于操作 HTTP 响应数据的方法，例如，设置响应状态码、设置响应头信息、发送响应数据到客户端等。HttpServletResponse 是一个接口，它还是 ServletResponse 的子接口。HttpServletResponse 接口常用的方法如表 3-8 所示。

表 3-8 HttpServletResponse 接口常用的方法

序号	方法名	描述
1	void addCookie(Cookie cookie)	向响应中添加一个 Cookie
2	void setHeader(String name, String value)	重置名称为 name 响应头的值为 value
3	void addHeader(String name, String value)	向响应中添加一个 name 响应头，其值为 value
4	String getHeader(String name)	返回名称为 name 的响应头的值
5	Collection\<String\> getHeaders(String name)	当响应头有多个值时，返回名字为 name 的响应头的值
6	Collection\<String\> getHeaderNames()	返回响应中所有响应头的名字
7	void setStatus(int sc)	设置响应的状态码
8	int getStatus()	返回响应的状态码
9	void sendError(int sc, String msg)	直接向客户端显示指定的错误码和错误描述
10	void sendRedirect(String location)	直接向客户端发送重定向消息，指示客户端重新发起对 location 地址的请求。使用这个方法可以完成客户端页面的跳转
11	String getCharacterEncoding()	返回响应的字符编码名称
12	void setCharacterEncoding(String charset)	设置响应体字符数据的编码类型
13	String getContentType()	返回响应体数据的 MIME 类型
14	void setContentType(String type)	设置响应体数据的 MIME 类型以及字符编码
15	void setContentLength(int len)	设置响应体中数据的长度，以字节为单位
16	ServletOutputStream getOutputStream()	返回一个可用于向客户端输出信息的数据流
17	PrintWriter getWriter()	返回一个可用于向客户端输出字符数据的打印数据流

3.5.1 发送响应数据到客户端

后端服务程序需要通过 HttpServletResponse 对象向前端服务程序发送响应数据。最常用的方式是从 HttpServletResponse 对象中获得一个 PrintWriter 对象，然后通过这个对象向前端服务程序发送字符数据，例如，发送 HTML 标签数据到前端程序。典型的代码如下：

```
resp.setContentType("text/html; charset=utf-8");
```

```
resp.getWriter().println("这是发送到客户端的信息");
```

一般而言,在使用 HttpServletResponse 对象发送结果数据到客户端程序之前,需要通过 setContentType()方法明确设置响应体的 MIME 类型和字符编码,否则,由于前端服务程序不了解响应的字符编码,会导致发生乱码现象。

后端服务程序除了可以通过 HttpServletResponse 对象向客户端发送字符数据外,也可以发送二进制类型的数据,例如,发送一张图片到客户端。下面编写一个称为 Image 的服务端程序,在浏览器中访问这个程序时将显示一张图片。Image.程序代码如下:

```
package com.ttt.servlet;
import jakarta.servlet.ServletException;
import jakarta.servlet.ServletOutputStream;
import jakarta.servlet.annotation.WebServlet;
import jakarta.servlet.http.HttpServlet;
import jakarta.servlet.http.HttpServletRequest;
import jakarta.servlet.http.HttpServletResponse;
import java.io.FileInputStream;
import java.io.IOException;
@WebServlet("/Image")
public class Image extends HttpServlet{
    protected void doGet(HttpServletRequest request, HttpServletResponse response)
        throws IOException{
        FileInputStream fis = new FileInputStream("C:\\Wu\\Temp\\wheat.jpg");
        byte[] b=new byte[fis.available()];
        fis.read(b);
        fis.close();
        response.setContentType("image/jpg");
        ServletOutputStream op = response.getOutputStream();
        op.write(b);
        op.close();
    }
    protected void doPost(HttpServletRequest request, HttpServletResponse response)
        throws ServletException, IOException{
        doGet(request, response);
    }
}
```

在这个程序中,首先使用第 1 组粗体字语句打开服务器本地磁盘的一张图片文件,并将图片文件内容读取到缓冲区 b 字节数组中。然后使用第 2 组粗体字语句设置响应体的 MIME 类型,并得到一个 ServletOutputStream 对象。最后通过这个输出流将图片数据发送到客户端。运行这个程序,在浏览器地址栏输入 Image 程序的 URL,运行结果如图 3-26 所示。

当然,也可以将 Image 程序的 URL 地址放在 HTML 页面中并用于显示图片。为此,在 ch03 工程的 html 目录下新建一个 image.html 页面文件,其内容如下:

```
<!DOCTYPE html>
<html lang="en">
<head>
    <meta charset="UTF-8">
    <title>显示图片</title>
</head>
<body>
```

图 3-26 后端服务程序直接返回二进制图片数据

```
    <h1>这是一张风景图片</h1>
    <img src="../Image" alt=""/>
</body>
</html>
```

运行程序,并在浏览器的地址栏输入 image.html,运行结果如图 3-27 所示。

图 3-27 从 HTML 页面访问 Image 服务程序

3.5.2　深入了解 setContentType 方法和 MIME

在从 HttpServletResponse 对象获取输出流之前,需要调用 setContentType 方法设置响应体的 MIME 类型。如果 MIME 类型为字符类型，还需要指定字符的编码类型。setContentType 方法的一般使用方式如下：

setContentType("MIME 类型；charset=字符集名称")

当 MIME 类型为文本数据时，需要使用"charset=字符集名称"指定内容字符的字符集名称；而当 MIME 类型为二进制数据时，可以省略"charset=字符集名称"。

在本书中已经多次提到 MIME 这个术语，那么，到底什么是 MIME？MIME 是 Multipurpose Internet Mail Extensions 的简称，是描述消息内容类型的标准，使用 MIME 可用来表示文档、文件或字节流的性质和格式。也就是说，当前端服务程序程序与后端服务程序要进行数据交换时，需要使用 MIME 来告诉对方所发送的数据的类型和格式。常用的 MIME 类型的名称及其含义如表 3-9 所示。

表 3-9 常用的 MIME 类型的名称及其含义

序号	MIME 类型名称	描述
1	application/msword	二进制数据，数据内容为微软 Office Word 格式
2	application/vnd.ms-excel	二进制数据，数据内容为微软 Office Excel 格式
3	application/vnd.ms-powerpoint	二进制数据，数据内容为微软 Office PPT 或 PPTX 格式
4	application/kswps	二进制数据，数据内容为 WPS 格式
5	application/pdf	二进制数据，数据内容为 PDF 格式
6	application/octet-stream	二进制数据，数据内容为二进制流数据
7	image/gif	二进制数据，数据内容为 GIF 图像
8	image/jpeg	二进制数据，数据内容为 JPEG 图像
9	image/png	二进制数据，数据内容为 PNG 图像
10	image/webp	二进制数据，数据内容为 WEBP 图像
11	text/plain	文本字符数据，数据内容为普通文本数据
12	text/javascript	文本字符数据，数据内容为 JavaScript 代码
13	text/css	文本字符数据，数据内容为 CSS 样式代码
14	text/html	文本字符数据，数据内容为 HTML 代码
15	audio/mpeg	二进制数据，数据内容为 MPEG 音频数据
16	audio/ogg	二进制数据，数据内容为 OGG 音频数据
17	video/mp4	二进制数据，数据内容为 MP4 视频数据
18	video/mpeg	二进制数据，数据内容为 MPEG 视频数据
19	video/webm	二进制数据，数据内容为 WEBM 视频数据
20	application/rar	二进制数据，数据内容为 RAR 压缩数据

3.5.3 发送状态码或错误信息

在 HTTP 的响应数据中有一个表示后端服务程序是否正确处理了前端服务程序请求的结果代码，这个代码称为状态码。HTTP 的状态码及其含义在第 2 章已经做了详细介绍。

默认情况下,HTTP响应的状态码被Servlet容器设置为200,表示后端服务程序正确处理了前端服务程序的请求。后端服务程序可以根据实际情况使用HttpServletResponse对象的setStatus()方法发送不同的状态码给前端服务程序,或者使用sendError()方法发送错误信息到前端服务程序。例如,新建一个名称为StatusTester的Servlet程序。StatusTester.java代码如下:

```java
package com.ttt.servlet;
import jakarta.servlet.ServletException;
import jakarta.servlet.annotation.WebServlet;
import jakarta.servlet.http.HttpServlet;
import jakarta.servlet.http.HttpServletRequest;
import jakarta.servlet.http.HttpServletResponse;
import java.io.IOException;
@WebServlet("/StatusTester")
public class StatusTester extends HttpServlet {
    @Override
    protected void doGet(HttpServletRequest req, HttpServletResponse resp)
        throws ServletException, IOException{
        resp.setContentType("text/html; charset=utf-8");
        String mess = "在客户端显示一些信息";
        resp.setStatus(300);
        resp.getWriter().println(mess);
    }
}
```

虽然后端服务程序正确处理了客户端的请求,但是,为了观察setStatus()方法的作用,使用这个方法发送了300这个状态码给客户端程序。运行这个程序,并在浏览器中访问StatusTester程序。运行结果如图3-28所示。

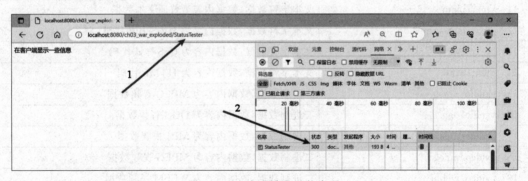

图3-28 向客户端发送状态码

打开浏览器的"开发者工具",注意箭头2所指向的信息:这就是服务程序设置的状态码。当然,服务程序所设置的状态码应该与HTTP规定的状态码的含义一致,才不会导致前端服务程序误解。

后端服务程序除了可以向前端服务程序发送指定的状态码外,还可以使用sendError()方法直接向前端服务程序发送错误消息。举一个简单例子。新建一个名称为ErrorTester的Servlet程序。ErrorTester.java代码如下:

```java
package com.ttt.servlet;
import jakarta.servlet.ServletException;
import jakarta.servlet.annotation.WebServlet;
import jakarta.servlet.http.HttpServlet;
import jakarta.servlet.http.HttpServletRequest;
import jakarta.servlet.http.HttpServletResponse;
import java.io.IOException;
@WebServlet("/ErrorTester")
public class ErrorTester extends HttpServlet{
    @Override
    protected void doGet(HttpServletRequest req, HttpServletResponse resp)
        throws ServletException, IOException{
        resp.setContentType("text/html; charset=utf-8");
        //由于 sendError()方法会清空输出流的信息,因此,下面的消息不会显示在客户端
        String mess = "在客户端显示一些信息";
        resp.setStatus(200);
        resp.getWriter().println(mess);
        //客户端只会显示这里的错误消息
        resp.sendError(500, "程序发送内部错误消息给客户端");
    }
}
```

运行这个程序,在浏览器地址栏输入 ErrorTester 程序的 URL 地址,显示如图 3-29 所示。

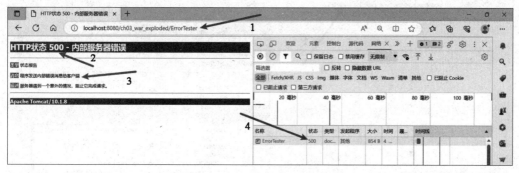

图 3-29 向客户端发送错误消息

图 3-29 所示的运行结果中,箭头 2 所指向的错误码和箭头 3 所指向的错误消息证实服务端程序使用 sendError()方法发送的。

3.5.4 设置响应消息头信息

后端服务程序可以通过 HTTP 响应头告知前端服务程序一些特殊信息。与 HTTP 请求头类似,HTTP 响应头也是"名:值"对形式。后端服务程序可以使用 HttpServletResponse 对象的 setHeader()方法和 addHeader()方法操作 HTTP 响应消息的响应头信息。举一个简单的例子。新建一个名为 Rheader 的 Servlet 程序。Rheader 程序代码如下:

```java
package com.ttt.servlet;
import jakarta.servlet.ServletException;
```

```java
import jakarta.servlet.annotation.WebServlet;
import jakarta.servlet.http.HttpServlet;
import jakarta.servlet.http.HttpServletRequest;
import jakarta.servlet.http.HttpServletResponse;
import java.io.IOException;
@WebServlet("/Rheader")
public class Rheader extends HttpServlet{
    @Override
    protected void doGet(HttpServletRequest req, HttpServletResponse resp)
            throws ServletException, IOException{
        resp.setContentType("text/html; charset=utf-8");
        resp.addHeader("user", "bill");
        resp.addHeader("user", "lisi");
        resp.setHeader("rank", "level 1");
        resp.getWriter().println("在HTTP相应中添加了新的响应头");
    }
}
```

在代码中，使用 addHeader()方法和 setHeader()方法设置了两个新的 HTTP 响应头。setHeader()和 addHeader()方法的区别是：addHeader()不会覆盖原有的头信息，而 setHeader()则会覆盖原有的头信息。运行这个程序，在浏览器地址栏中输入 Rheader 程序的 URL 地址，显示如图 3-30 所示的结果。

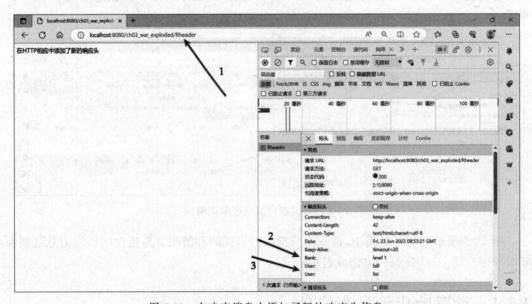

图 3-30 在响应消息中添加了新的响应头信息

如图 3-30 所示，在 HTTP 的响应消息头中多了服务端程序新增的 user 和 rank 头信息。

在对 HTTP 的响应头进行操作时存在一个限制：不能将中文字符作为响应头 value 值。如果需要将中文字符作为响应头的 value，则必须进行编码，如下代码所示：

```
String data = URLEncoder.encode("张三", StandardCharsets.UTF_8);
resp.addHeader("user", data);
```

在读取进行编码的响应头时,必须进行解码,如下代码所示:

```
String data = resp.getHeader("user");
data = URLDecoder.decode(data, "utf-8");
```

3.5.5 请求重定向

所谓"请求重定向",是指后端服务程序通过 HttpServletResponse 对象告知前端服务程序重新发起一个对指定的新 URL 的 HTTP 请求。HttpServletResponse 对象的 sendRedirect()方法即可完成请求重定向功能。请求重定向工作过程如图 3-31 所示。

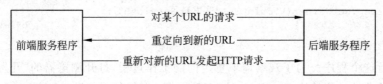

图 3-31 请求重定向工作过程

下面举例说明请求重定向的使用。在这个例子中,前端服务程序请求后端服务程序 ServletA 提供服务,但是 ServletA 直接将前端服务程序的这个请求重定向到 ServletB。 ServletA 的代码如下:

```
package com.ttt.servlet;
import jakarta.servlet.ServletException;
import jakarta.servlet.annotation.WebServlet;
import jakarta.servlet.http.HttpServlet;
import jakarta.servlet.http.HttpServletRequest;
import jakarta.servlet.http.HttpServletResponse;
import java.io.IOException;
@WebServlet("/ServletA")
public class ServletA extends HttpServlet{
    @Override
    protected void doGet(HttpServletRequest req, HttpServletResponse resp)
            throws ServletException, IOException{
        resp.setContentType("text/html; charset=UTF-8");
        //以下信息是不会在前端服务程序中显示的
        resp.getWriter().println("重定向到新的 URL,这里是 ServletB");

        resp.sendRedirect("./ServletB");
    }
}
```

在 ServletA 中,使用语句"resp.getWriter().println("重定向到新的 URL,这里是 ServletB");",期望在前端服务程序中显示一段文字信息。但是,由于后面使用语句"resp.sendRedirect("./ServletB");"做了请求重定向,因此,这段信息是不会在前端服务程序中显示出来。通过重定向,将前端服务程序的请求直接导向到了"./ServletB"。ServletB 的代码如下:

```
package com.ttt.servlet;
```

```
import jakarta.servlet.ServletException;
import jakarta.servlet.annotation.WebServlet;
import jakarta.servlet.http.HttpServlet;
import jakarta.servlet.http.HttpServletRequest;
import jakarta.servlet.http.HttpServletResponse;
import java.io.IOException;
@WebServlet("/ServletB")
public class ServletB extends HttpServlet{
    @Override
    protected void doGet(HttpServletRequest req, HttpServletResponse resp)
            throws ServletException, IOException{
        resp.setContentType("text/html; charset=UTF-8");
        resp.getWriter().println("这里是ServletB");
    }
}
```

现在运行这个程序。为了观察请求重定向的工作过程，打开浏览器的"开发人员工具"，然后在浏览器的地址栏输入 ServletA 的 URL 地址，运行结果如图 3-32 所示。

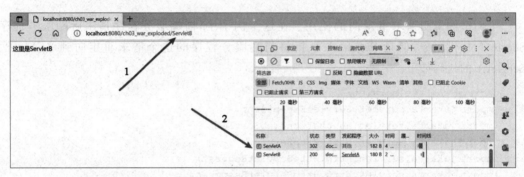

图 3-32　请求重定向的工作过程

从图 3-32 中可以发现，虽然在浏览器地址栏输入了 ServletA 的 URL 地址，但是，按下 Enter 键，浏览器地址栏显示的却是 ServletB 的 URL 地址，这是为什么呢？观察图 3-32 箭头 2 所指向的信息可以发现，浏览器向服务器发送了两个请求：第一个请求是请求 ServletA 提供服务；第二个则是请求 ServletB 提供服务。这是由于当浏览器请求 ServletA 提供服务时，ServletA 直接告诉浏览器向 ServletB 发请求，浏览器则再次向 ServletB 发送请求，所以，浏览器地址栏显示的是 ServletB 的地址。这也正体现了请求重定向的工作过程。

3.6　ServletContext 对象及其使用

把 Java Web 应用程序部署到 Servlet 容器时，Servlet 容器会为每个 Java Web 程序创建一个全局唯一的称为 ServletContext 的对象。例如，当把 ch03 程序部署到 Tomcat 时，Tomcat 会为 ch03 程序创建全局唯一的 ServletContext 对象。

ServletContext 对象提供了一系列的方法，使得 Servlet 程序与 Servlet 容器可以进行信息交互。任何服务端 Servlet 程序可以通过调用 getServletContext()方法获得这个全局共享的 ServletContext 对象。本质上，ServletContext 是一个接口，Servlet 容器会实例化 ServletContext 接口对象。ServletContext 接口常用的方法如表 3-10 所示。

表 3-10　ServletContext 接口常用的方法

序号	方 法 名	描 述
1	Object getAttribute(String name)	获取名称为 name 的属性的值
2	Enumeration<String> getAttributeNames()	获取所有属性的名字
3	void removeAttribute(String name)	删除名为 name 的属性
4	void setAttribute(String name, Object object)	设置 name 属性的值为 object 对象
5	StringgetContextPath()	返回应用程序的上下文路径。例如，对于 ch03 程序，其部署的上下文路径为 "/ch03_war_exploded"
6	String getRealPath(String path)	返回 path 路径的真实路径。例如对于 ch03 程序，调用 getRealPath("/aa")，则返回 "C:\Wu\javaweb\out\artifacts\ch03_war_exploded\aa"
7	int getMajorVersion()	返回 Servlet 容器支持的 Servlet 主版本号
8	int getMinorVersion()	返回 Servlet 容器支持的 Servlet 子版本号
9	String getServerInfo()	返回 Servlet 容器的名称和版本信息
10	InputStream getResourceAsStream(String path)	得到指定资源路径的 Input Stream 流对象
11	void log(String msg)	写一条日志信息到容器的日志文件中，功能类似于 Servlet 接口的同名方法
12	String getRequestCharacterEncoding()	返回 Servlet 容器默认的请求字符编码
13	String getResponseCharacterEncoding()	返回 Servlet 容器默认的响应字符编码
14	void setRequestCharacterEncoding(String encoding)	设置 Servlet 容器默认的请求字符编码
15	void setResponseCharacterEncoding(String encoding)	设置 Servlet 容器默认的响应字符编码
16	String getInitParameter(String name)	返回配置文件中给定的 name 参数的值
17	Enumeration<String> getInitParameterNames()	返回配置文件中所有配置参数的名字
18	int getSessionTimeout()	返回会话超时时间，单位是分钟。关于会话技术，将在第 5 章介绍
19	void setSessionTimeout(int sessionTimeout)	设置 Servlet 容器会话的超时时间，单位是分钟

ServletContext 接口的最常用方式是实现 Servlet 程序之间的信息共享和使用 ServletContext 接口读取资源文件。

3.6.1　使用 ServletContext 实现数据共享

ServletContext 接口最常用的方式是使 Servlet 之间实现数据共享，例如，一个 Servlet 程序通过调用 ServletContext 接口的 setAttribute()方法设置某个属性，另一个 Servlet 可以通过调用 ServletContext 接口的 getAttribute()方法获取属性值。下面通过一个例子介绍如何使用 ServletContext 实现信息共享。

在这个例子中有两个 Servlet 程序:ServletS1 和 ServletS2。当 ServletS1 被访问时,它会在 ServletContext 对象中保存一个名字为 date 的属性,用于记录该 Servlet 最近被访问的日期时间;当 ServletS2 被访问时,ServletS2 会从 ServletContext 对象读取名字为 date 的属性值并在浏览器中显示出来。ServletS1.java 程序代码如下:

```java
package com.ttt.servlet;
import jakarta.servlet.ServletContext;
import jakarta.servlet.ServletException;
import jakarta.servlet.annotation.WebServlet;
import jakarta.servlet.http.HttpServlet;
import jakarta.servlet.http.HttpServletRequest;
import jakarta.servlet.http.HttpServletResponse;
import java.io.IOException;
import java.util.Date;
@WebServlet("/ServletS1")
public class ServletS1 extends HttpServlet {
    @Override
    protected void doGet(HttpServletRequest req, HttpServletResponse resp)
            throws ServletException, IOException{
        ServletContext sc = getServletContext();
        sc.setAttribute("date", new Date());
        resp.setContentType("text/html; charset=utf-8");
        resp.getWriter().println("已经保存了最新访问日期到 ServletContext 对象");
    }
}
```

在 ServletS1 程序中,首先获得 ServletContext 对象,然后在这个对象中保存了 ServletS1 最近被访问的日期时间。ServletS2.java 程序如下所述。

```java
package com.ttt.servlet;
import jakarta.servlet.ServletContext;
import jakarta.servlet.ServletException;
import jakarta.servlet.annotation.WebServlet;
import jakarta.servlet.http.HttpServlet;
import jakarta.servlet.http.HttpServletRequest;
import jakarta.servlet.http.HttpServletResponse;
import java.io.IOException;
import java.text.SimpleDateFormat;
import java.util.Date;
@WebServlet("/ServletS2")
public class ServletS2 extends HttpServlet {
    @Override
    protected void doGet(HttpServletRequest req, HttpServletResponse resp)
            throws ServletException, IOException{
        ServletContext sc = getServletContext();
        Date d = (Date)sc.getAttribute("date");
        String info;
        if (d == null){
            info = "最近 ServletS1 没有被访问";
        }
        else{
```

```
        SimpleDateFormat sdf = new SimpleDateFormat("yyyy-MM-dd HH:mm:ss");
        info = sdf.format(d);
    }
    resp.setContentType("text/html;charset=utf-8");
    resp.getWriter().println(info);
}
```

在 ServletS2 程序中，首先获得 ServletContext 对象，然后从中获取属性名为 date 属性的值，并根据是否存在这个属性在浏览器中显示不同的结果。注意，为了使日期结果符合中国人习惯，程序中使用了 SimpleDateFormat 对象对日期格式做了规范化操作。现在运行这个程序：如果在没有访问 ServletS1 的情况下先访问 ServletS2，显示如图 3-33 所示的结果。

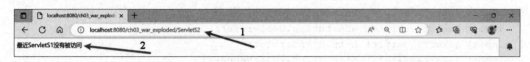

图 3-33　在没有访问 ServletA 的情况下先访问 ServletB 的运行结果

因为没有访问过 ServletS1，所以在 ServletContext 中不存在 date 属性，因此，显示的结果应该如此。现在访问 ServletS1，之后再访问 ServletS2，则运行结果如图 3-34 所示。

图 3-34　在访问 ServletA 的之后再访问 ServletB 的运行结果

虽然程序之间通过 ServletContext 实现了属性数据的共享，但是，目前的程序存在一个漏洞。设想一下，如果同时有 10000 个人同时访问 ServletS1，由于线程的分时运行，可能出现这样的情况：一个 ServletS1 的线程正在向 ServletContext 对象存在自己被访问的时间数据，由于时间片到时，操作系统会剥夺这个线程的运行权利而切换到另一个线程运行，这样势必导致在 ServletContext 对象所存在的属性数据的一致性出现问题。为了解决这个问题，需要对 ServletContext 的访问进行必要的保护。Java 提供了数据保护措施：sychronized 关键字是最简单的方式。为此，分别修改 ServletS1 和 ServletS2 为 ServletS3 和 ServletS4。ServletS3 的代码如下：

```
package com.ttt.servlet;
import jakarta.servlet.ServletContext;
import jakarta.servlet.ServletException;
import jakarta.servlet.annotation.WebServlet;
import jakarta.servlet.http.HttpServlet;
import jakarta.servlet.http.HttpServletRequest;
import jakarta.servlet.http.HttpServletResponse;
import java.io.IOException;
import java.util.Date;
@WebServlet("/ServletS3")
public class ServletS3 extends HttpServlet {
    @Override
    protected void doGet(HttpServletRequest req, HttpServletResponse resp)
```

```java
    throws ServletException, IOException {
        ServletContext sc = getServletContext();
        synchronized (sc){
            sc.setAttribute("date", new Date());
        }
        resp.setContentType("text/html; charset=utf-8");
        resp.getWriter().println("已经保存了最新访问日期到 ServletContext 对象");
    }
}
```

ServletS4 的代码如下：

```java
package com.ttt.servlet;
import jakarta.servlet.ServletContext;
import jakarta.servlet.ServletException;
import jakarta.servlet.annotation.WebServlet;
import jakarta.servlet.http.HttpServlet;
import jakarta.servlet.http.HttpServletRequest;
import jakarta.servlet.http.HttpServletResponse;
import java.io.IOException;
import java.text.SimpleDateFormat;
import java.util.Date;
@WebServlet("/ServletS4")
public classServletS4 extends HttpServlet {
    @Override
    protected void doGet(HttpServletRequest req, HttpServletResponse resp)
        throws ServletException, IOException{
        ServletContext sc = getServletContext();
        Date d = null;
        synchronized (sc) {
            d = (Date) sc.getAttribute("date");
        }
        String info;
        if (d == null){
            info = "最近 ServletS3 没有被访问";
        }
        else{
            SimpleDateFormat sdf = new SimpleDateFormat("yyyy-MM-dd HH:mm:ss");
            info = sdf.format(d);
        }
        resp.setContentType("text/html; charset=utf-8");
        resp.getWriter().println(info);
    }
}
```

在设计服务端程序时，如果线程之间存在数据共享，需要特别注意并发操作时对数据的保护，否则可能由于数据的一致性问题而发生不可预料的程序缺陷。

3.6.2 使用 ServletContext 读取资源文件

ServletContext 另一个重要用途是读取资源文件。在新建 Java Web 程序工程时，

IDEA 会为每个工程创建用于存放程序源代码和程序资源文件的工程结构。例如，对 ch03 工程，其目录结构如图 3-35 所示。

图 3-35 工程/模块的工程结构

在图 3-35 中，箭头 1 所指向的 java 目录用于存放 Java 源代码，而箭头 2 所指向的 resources 则用于存放程序资源文件，例如，程序的配置文件等。现在，在 resources 资源目录下新建一个 config.properties 资源文件。config.properties 文件的内容如下：

```
username=bill3000
level=analysis
language=java
version=17
```

现在部署 ch03 程序，然后打开部署目录 target，如图 3-36 所示。

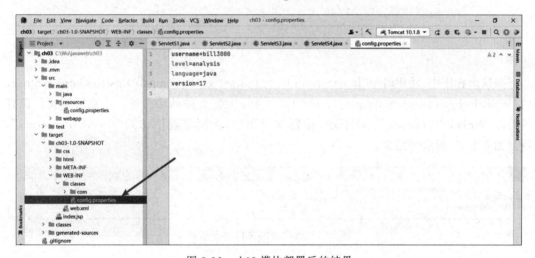

图 3-36 ch03 模块部署后的结果

从图 3-36 中可以看出，程序部署后，源代码 java 目录各个包的 class 文件会出现在上下文路径的 WEB-INF/classes 目录下，并以各个类的包名为目录进行存储；而在 resources 资源目录下的文件也会出现在上下文路径的 WEB-INF/classes 目录下。基于此，使用 ServletContext 对象的 getResourceAsStream()方法就可以访问到相应的资源文件。下面举一个例子来读取 config.properties 配置文件的内容。为此，新建 ServletProperties 服务端 Servlet 程序，其代码如下：

```
package com.ttt.servlet;
import jakarta.servlet.ServletContext;
import jakarta.servlet.ServletException;
```

```java
import jakarta.servlet.annotation.WebServlet;
import jakarta.servlet.http.HttpServlet;
import jakarta.servlet.http.HttpServletRequest;
import jakarta.servlet.http.HttpServletResponse;
import java.io.IOException;
import java.io.InputStream;
import java.util.Properties;
@WebServlet("/ServletProperties")
public class ServletProperties extends HttpServlet{
    @Override
    protected void doGet(HttpServletRequest req, HttpServletResponse resp)
      throws ServletException, IOException{
        ServletContext sc = getServletContext();
        InputStream is = this.getServletContext().
            getResourceAsStream("/WEB-INF/classes/config.properties");
        Properties prop = new Properties();
        prop.load(is);
        String username = prop.getProperty("username");
        String level = prop.getProperty("level");
        String language = prop.getProperty("language");
        String version = prop.getProperty("version");
        resp.setContentType("text/html; charset=utf-8");
        resp.getWriter().println(username + " : " + level + " : " +
            language + " : " + version);
    }
}
```

在这个程序中,使用语句"InputStream is = this.getServletContext().getResourceAsStream("/WEB-INF/classes/config.properties");"创建了配置文件的输入流。注意资源文件路径的是以/WEB-INF/classes 为父目录。运行这个程序,在浏览器中访问"/ServletProperties",得到如图 3-37 所示的结果。

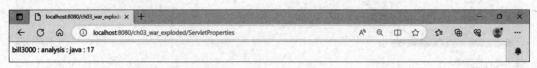

图 3-37　读取资源文件运行结果

从图 3-37 中可以看出,ServletProperties 正确读取了 config.properties 文件内容,并正确显示在浏览器里。

当然,如果程序的配置文件比较多,也可以根据需要在 resources 目录下创建子目录,然后在子目录下放置资源文件:在 resources 目录下的所有子目录及其文件,都将原封不动地复制到 WEB-INF/classes 部署目录下。

3.6.3　关于 web.xml 配置文件

任何一个 Java Web 工程中都存在一个名为 web.xml 的配置文件,通过这个 web.xml 文件可以对 Java Web 程序进行业务配置,包括:对 Servlet 的配置,类似于@WebServlet 注解,可以对 Servlet 的 URL 映射配置、初始参数等进行配置等;对应用程序启动参数的配置;对 Servlet 加载顺序的配置等。

关于 web.xml 配置文件

3.7 案例:用户注册

下面编写一个用户注册程序来结束本章的学习。

3.7.1 案例目标

编写程序实现用户注册功能。

编写一个 Servlet 程序,接收用户在浏览器上输入的用户注册信息,包括用户名、密码、电话号码等基本信息。当用户单击页面的"注册"按钮后,将用户注册信息发送到后台 Servlet 并保存在后台服务器的日志中,同时,将用户的注册信息发送到前端浏览器中显示出来。

3.7.2 案例分析

需要编写一个前端页面程序,用户可以在页面上输入包括用户名、密码、电话号码等基本信息,然后单击"提交"按钮将信息提交给服务端程序进行处理。页面必须是美观的,因此,需要编写页面布局 CSS 程序并用于美化页面。为了完成这个程序的功能目标,需要编写以下几个程序:

(1) 页面 HTML 程序,命名这个程序文件为 register.html;
(2) 页面布局 CSS 程序,命名这个程序文件为 register.css;
(3) 后端服务处理程序,命名这个程序文件为 Register.java。

3.7.3 案例实施

在 ch03 工程的"/webapp/html"目录下新建名称为 register.html 的页面文件,该文件的内容如下:

```
<!DOCTYPE html>
<html lang="js">
<head>
  <meta charset="utf-8">
```

```html
    <title>用户注册</title>
    <link href="../css/register.css" rel="stylesheet" type="text/css" />
</head>
<body>
<h3>请填写注册信息</h3>
<div>
    <form enctype="application/x-www-form-urlencoded"
                    method="POST" action="../Register">
        <label for="username">用户名</label>
        <input type="text" id="username" name="username" placeholder="用户名...">
        <label for="password1">密码</label>
        <input type="password" id="password1" name="password1" placeholder="密码...">
        <label for="password2">密码确认</label>
        <input type="password" id="password2" name="password2" placeholder="密码确认...">
        <label for="phone">电话号码</label>
        <input type="text" id="phone" name="phone" placeholder="电话号码...">
        <p>性别:
            <input type="radio" value="boy" name="gender" checked/>男
            <input type="radio" value="girl" name="gender"/>女
        </p>
        <p>爱好:
            <input type="checkbox" value="reading" name="hobby" checked/>读书
            <input type="checkbox" value="running" name="hobby"/>跑步
            <input type="checkbox" value="chatting" name="hobby"/>聊天
            <input type="checkbox" value="traveling" name="hobby"/>旅游
        </p>
        <input type="submit" value="提交">
    </form>
</div>
</body>
</html>
```

相应地,在 ch03 工程的"/webapp/css"目录下新建名称为 register.css 的样式布局文件,该文件的内容如下:

```css
input[type=text], input[type=password], p, select{
    width: 100%;
    padding: 12px 20px;
    margin: 8px 0;
    display: inline-block;
    border: 1px solid #ccc;
    border-radius: 4px;
    box-sizing: border-box;
}
input[type=submit]{
    width: 100%;
    background-color: #4CAF50;
    color: white;
    padding: 14px 20px;
    margin: 8px0;
    border: none;
```

```css
    border-radius: 4px;
    cursor: pointer;
}
input[type=submit]:hover{
    background-color: #45a049;
}
div{
    border-radius: 5px;
    background-color: #f2f2f2;
    padding: 20px;
}
```

最后，在 ch03 工程中新建名称为 Register 的 Servlet 程序，Register.java 程序代码如下：

```java
package com.ttt.servlet;
import jakarta.servlet.ServletException;
import jakarta.servlet.annotation.WebServlet;
import jakarta.servlet.http.HttpServlet;
import jakarta.servlet.http.HttpServletRequest;
import jakarta.servlet.http.HttpServletResponse;
import java.io.IOException;
@WebServlet("/Register")
public class Register extends HttpServlet{
    @Override
    protected void doPost(HttpServletRequest req, HttpServletResponse resp)
      throws ServletException, IOException{
        req.setCharacterEncoding("utf-8");
        String username = req.getParameter("username");
        String password1 = req.getParameter("password1");
        String password2 = req.getParameter("password2");
        String phone = req.getParameter("phone");
        String gender = req.getParameter("gender");
        String[] hobby = req.getParameterValues("hobby");
        resp.setContentType("text/html; charset=utf-8");
        StringBuffer sb = new StringBuffer();
        if (username.isBlank())
            sb.append("必须输入用户名<br>");
        else if (password1.isBlank())
            sb.append("必须输入密码<br>");
        else if (password2.isBlank())
            sb.append("必须输入密码<br>");
        else if (!password1.equals(password2))
            sb.append("两次输入的密码不一致<br>");
        else if (phone.isBlank())
            sb.append("必须输入电话号码<br>");
        else if (gender.isBlank())
            sb.append("必须选择性别<br>");
        else if (hobby.length == 0)
         sb.append("必须选择爱好<br>");
        if (sb.length() != 0) {
```

```
            resp.getWriter().println(sb);
            return;
        }
        sb.append("用户名: ").append(username).append("<br>");
        sb.append("密码: ").append(password1).append("<br>");
        sb.append("电话号码: ").append(phone).append("<br>");
        sb.append("性别: ").append(gender).append("<br>");
        sb.append("爱好: ");
        for(String h:hobby){
            sb.append(h).append("  ");
        }
        sb.append("<br>");
        resp.getWriter().println(sb);
        log(sb.toString());
    }
}
```

在这个程序中,使用如下语句:

```
String username = req.getParameter("username");
String password1 = req.getParameter("password1");
String password2 = req.getParameter("password2");
String phone = req.getParameter("phone");
String gender = req.getParameter("gender");
String[] hobby = req.getParameterValues("hobby");
```

从请求中获取用户在页面中输入的信息。注意语句:

String[] hobby = req.getParameterValues("hobby");

由于用户可以选择多个爱好,所以使用 getParameterValues 获取用户的多个选择。运行这个程序,在浏览器地址栏输入 register.html 页面,显示如图 3-38 所示的界面。

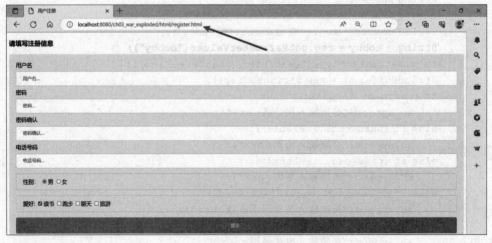

图 3-38 用户注册页面

在注册页面中输入相应信息,然后单击"提交"按钮,将显示如图 3-39 所示界面。

此时打开 Tomcat 的日志文件,会发现在日志文件中也记录了相应信息,如图 3-40 所示。

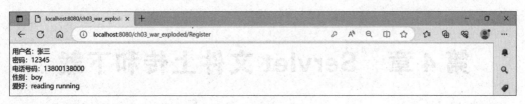

图 3-39 提交注册结果页面

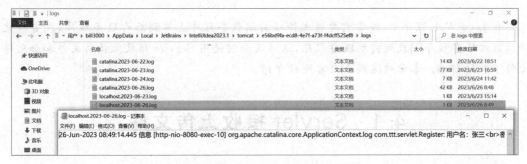

图 3-40 日志文件内容

3.8 练习：编写书籍录入程序

编写书籍信息录入程序。要求首先制作一个书籍录入页面，可以录入书籍的基本信息，包括书籍名称、出版社、编者、价格、书籍介绍，当用户在页面上录入信息并单击"提交"按钮后，将书籍信息保存在后台服务器中。

第 4 章 Servlet 文件上传和下载

在 Java Web 程序中,经常需要将文件从前端服务程序上传到后台服务程序,或者将文件从后端服务程序下载到前端服务程序,这涉及如何使用 Servlet 接收上传的文件和如何将文件下载到前端。本章对这两个内容进行介绍。

4.1 Servlet 接收上传文件

所谓"Servlet 接收上传文件"是指 Servlet 服务程序接收从前端程序上传的文件。上传的文件类型可以是任何格式,包括图片文件、PDF 文件、Word 文件等。作为例子,下面完善第 3 章的用户注册程序:除了包括用户的基本信息外,还要求用户上传一张个人头像。为此,新建一个名为 ch04 的 Java Web 工程,并将第 3 章的 register.html 文件、register.css 文件和 Register.java 文件复制到 ch04 工程的相应目录下。新建完成的 ch04 工程结构如图 4-1 所示。

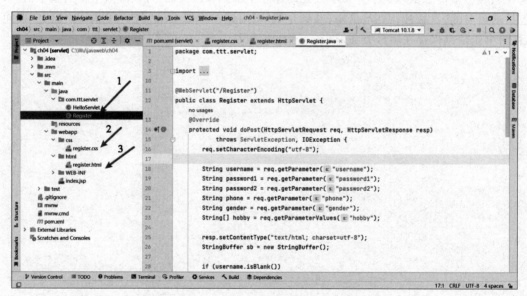

图 4-1 新建完成的 ch04 工程结构

总体来说,上传文件到服务端包括两个关键要素:其一,在前端程序中设计一个包含文件上传的 form 表单;其二,在后端服务程序中利用 @MultipartConfig 注解标注接收文件的 Servlet 程序并利用 Part 类处理文件。

4.1.1 编写包含上传文件功能的注册页面

要使 Servlet 服务程序能够接收文件,前端程序必须要首先上传文件到服务端。因此,首先修改前端程序的页面代码 register.html 文件,使之可以上传文件到服务端的 Register 程序。修改后的 register.html 程序如下:

```html
<!DOCTYPE html>
<html lang="js">
<head>
  <meta charset="utf-8">
  <title>用户注册</title>
  <link href="../css/register.css" rel="stylesheet" type="text/css" />
</head>
<body>
<h3>请填写注册信息</h3>
<div>
  <form enctype="multipart/form-data" method="POST" action="../Register">
    <label for="username">用户名</label>
    <input type="text" id="username" name="username" placeholder="用户名...">
    <label for="password1">密码</label>
    <input type="password" id="password1" name="password1" placeholder="密码...">
    <label for="password2">密码确认</label>
    <input type="password" id="password2" name="password2" placeholder="密码确认...">
    <label for="phone">电话号码</label>
    <input type="text" id="phone" name="phone" placeholder="电话号码...">
    <p>性别:
      <input type="radio" value="boy" name="gender" checked/>男
      <input type="radio" value="girl" name="gender"/>女
    </p>
    <p>爱好:
      <input type="checkbox" value="reading" name="hobby" checked/>读书
      <input type="checkbox" value="running" name="hobby"/>跑步
      <input type="checkbox" value="chatting" name="hobby"/>聊天
      <input type="checkbox" value="traveling" name="hobby"/>旅游
    </p>
    <label for="photo">头像</label>
    <input type="file" id="photo" name="photo" placeholder="选择头像文件...">
    <input type="submit" value="提交">
  </form>
</div>
</body>
</html>
```

在 register.html 的代码中,为了在 form 表单中上传文件,需要将表单的封装格式设置为 multipart/form-data,且请求方法必须为 POST 方法,代码如下所示:

```html
<form enctype="multipart/form-data" method="POST" action="../Register">
```

同时,在表单中增加了如下选择头像文件的标签:

```
<label for="photo">头像</label>
<input type="file" id="photo" name="photo" placeholder="选择头像文件...">
```

为了使页面美观,也修改了 register.css 样式文件,修改后的 register.css 文件内容如下:

```css
input[type=text], input[type=password], input[type=file], p, select {
    width: 100%;
    padding: 12px 20px;
    margin: 8px 0;
    display: inline-block;
    border: 1px solid #ccc;
    border-radius: 4px;
    box-sizing: border-box;
}
input[type=submit]{
    width: 100%;
    background-color: #4CAF50;
    color: white;
    padding: 14px 20px;
    margin: 8px 0;
    border: none;
    border-radius: 4px;
    cursor: pointer;
}
input[type=submit]:hover{
    background-color: #45a049;
}
div{
    border-radius: 5px;
    background-color: #f2f2f2;
    padding: 20px;
}
```

注册页面中包含了文件选择输入框,现在需要修改服务端 Register 程序接收客户端发送的注册信息。

4.1.2 接收客户端上传的头像文件

为了接收从客户端上传的文件,Java EE 定义了 @MultipartConfig 注解和与文件上传相关的类。@MultipartConfig 注解会分析 HTTP 的 multipart/form-data 请求,根据名称从请求中获取指定的内容,程序可对获取得到的内容信息进行进一步处理。现在,修改 Register 服务端代码,修改后的 Register.java 代码如下:

```java
package com.ttt.servlet;
import jakarta.servlet.ServletException;
import jakarta.servlet.annotation.MultipartConfig;
import jakarta.servlet.annotation.WebServlet;
import jakarta.servlet.http.HttpServlet;
import jakarta.servlet.http.HttpServletRequest;
```

```java
import jakarta.servlet.http.HttpServletResponse;
import jakarta.servlet.http.Part;
import java.io.ByteArrayOutputStream;
import java.io.FileOutputStream;
import java.io.IOException;
import java.io.InputStream;
import java.util.UUID;
@MultipartConfig
@WebServlet("/Register")
public class Register extends HttpServlet{
    @Override
    protected void doPost(HttpServletRequest req, HttpServletResponse resp)
            throws ServletException, IOException{
        req.setCharacterEncoding("utf-8");
        String username = req.getParameter("username");
        String password1 = req.getParameter("password1");
        String password2 = req.getParameter("password2");
        String phone = req.getParameter("phone");
        String gender = req.getParameter("gender");
        String[] hobby = req.getParameterValues("hobby");
        Part photo = req.getPart("photo");
        resp.setContentType("text/html; charset=utf-8");
        StringBuffer sb = new StringBuffer();
        if (username.isBlank())
            sb.append("必须输入用户名<br>");
        else if (password1.isBlank())
            sb.append("必须输入密码<br>");
        else if (password2.isBlank())
            sb.append("必须输入密码<br>");
        else if (!password1.equals(password2))
            sb.append("两次输入的密码不一致<br>");
        else if (phone.isBlank())
            sb.append("必须输入电话号码<br>");
        else if (gender.isBlank())
            sb.append("必须选择性别<br>");
        else if (hobby.length == 0)
            sb.append("必须选择爱好<br>");
        else if (photo == null)
            sb.append("必须上传头像");
        if (sb.length() != 0) {
            resp.getWriter().println(sb);
            return;
        }
        sb.append("用户名：").append(username).append("<br>");
        sb.append("密码：").append(password1).append("<br>");
        sb.append("电话号码：").append(phone).append("<br>");
        sb.append("性别：").append(gender).append("<br>");
        sb.append("爱好：");
        for(String h:hobby){
            sb.append(h).append("   ");
        }
```

```
            sb.append("<br>");
            assert photo != null;
            InputStream is = photo.getInputStream();
            ByteArrayOutputStream baos = new ByteArrayOutputStream();
            byte[] b = new byte[1024];
            while(is.read(b)>0) {
                baos.write(b);
            }
            b = baos.toByteArray();
            UUID uuid = UUID.randomUUID();
            String sfn = photo.getSubmittedFileName();
            String suffix = sfn.substring(sfn.lastIndexOf("."));
            FileOutputStream fos = newFileOutputStream("C:/photo/"+
                    uuid.toString()+suffix);
            fos.write(b);
            fos.close();
            resp.getWriter().println(sb);
        }
    }
```

在 Register.java 程序代码中,语句@MultipartConfig 告知 Servlet 容器对该服务端程序的 HTTP 请求进行分析,然后使用语句"Part photo = req.getPart("photo");"从请求中获取名称为 photo 的部分。注意,这里的名称必须与 form 表单中的名称一致。然后,使用如下语句:

```
InputStream is = photo.getInputStream();
ByteArrayOutputStream baos = new ByteArrayOutputStream();
byte[] b = new byte[1024];
while(is.read(b)>0) {
    baos.write(b);
}
b = baos.toByteArray();
```

从中得到一个输入流,并将该输入流中的字节存入字节数组流中。最后,使用如下语句:

```
UUID uuid = UUID.randomUUID();
String sfn = photo.getSubmittedFileName();
String suffix = sfn.substring(sfn.lastIndexOf("."));
FileOutputStream fos = new FileOutputStream("C:/photo/"+
                    uuid.toString()+suffix);
fos.write(b);
fos.close();
```

在利用 UUID 生成一个唯一的文件名后,再将字节流中的数据保存到 C:/photo 目录下指定的文件中。

现在,运行这个程序,在浏览器地址栏输入 register.html 页面地址,显示如图 4-2 所示的界面。

在图 4-2 的界面中输入相应信息并选择头像文件,然后单击"提交"按钮,注册信息被提交到 Register 服务端程序处理后,将显示如图 4-3 所示的界面。

第 4 章　Servlet 文件上传和下载

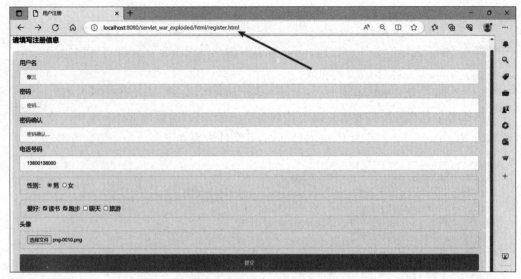

图 4-2　包含上传文件的注册页面

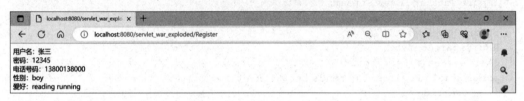

图 4-3　注册成功后的信息

如果打开 C:/photo 目录,会发现里面保存了客户端上传的文件,打开这个文件会发现,这个文件正是上传的头像文件。

4.1.3　多文件上传

有时也需要一次上传多个文件到服务端。form 表单支持多文件上传,只需在上传文件的 input 标签中增加 multiple 即可。例如,在用户注册页面中,如果需要同时上传多个文件,只需简单地对 register.html 修改即可。命名修改后的 register.html 文件为 register1.html,其文件内容如下:

```
//与register.html代码一致,此处省略了部分代码
<body>
<h3>请填写注册信息</h3>
<div>
  <form enctype="multipart/form-data" method="POST" action="../Register1">
    <label for="username">用户名</label>
    <input type="text" id="username" name="username" placeholder="用户名...">
    //与register.html代码一致,此处省略了部分代码
    <label for="photo">头像</label>
    <input type="file" multiple id="photo" name="photo" placeholder="选择头像文件...">
    <input type="submit" value="提交">
  </form>
```

```
</div>
//与register.html代码一致,此处省略了部分代码
```

只需在form表单的文件中选择input标签并增加multiple属性即可,语句如下:

```
<label for="photo">头像</label>
<input type="file" multiple id="photo" name="photo" placeholder="选择头像文件...">
```

当然服务端程序需要接收客户端上传的多个文件。为此,修改Register.java文件,命名修改后的Register.java文件为Register1.java,其文件内容如下:

```
package com.ttt.servlet;
import jakarta.servlet.ServletException;
import jakarta.servlet.annotation.MultipartConfig;
import jakarta.servlet.annotation.WebServlet;
import jakarta.servlet.http.HttpServlet;
import jakarta.servlet.http.HttpServletRequest;
import jakarta.servlet.http.HttpServletResponse;
import jakarta.servlet.http.Part;
import java.io.IOException;
import java.io.InputStream;
import java.util.Collection;
import java.util.UUID;
@MultipartConfig
@WebServlet("/Register1")
public class Register1 extends HttpServlet{
    @Override
    protected void doPost(HttpServletRequest req, HttpServletResponse resp)
            throws ServletException, IOException{
        req.setCharacterEncoding("utf-8");
        String username = req.getParameter("username");
        String password1 = req.getParameter("password1");
        String password2 = req.getParameter("password2");
        String phone = req.getParameter("phone");
        String gender = req.getParameter("gender");
        String[] hobby = req.getParameterValues("hobby");
        resp.setContentType("text/html; charset=utf-8");
        StringBuffer sb = new StringBuffer();
        if (username.isBlank())
            sb.append("必须输入用户名<br>");
        else if (password1.isBlank())
            sb.append("必须输入密码<br>");
        else if (password2.isBlank())
            sb.append("必须输入密码<br>");
        else if (!password1.equals(password2))
            sb.append("两次输入的密码不一致<br>");
        else if (phone.isBlank())
            sb.append("必须输入电话号码<br>");
        else if (gender.isBlank())
            sb.append("必须选择性别<br>");
        else if (hobby.length == 0)
            sb.append("必须选择爱好<br>");
        if (sb.length() != 0) {
            resp.getWriter().println(sb);
            return;
```

```
        }
        Collection<Part> parts = req.getParts();
        for(Part p : parts){
            if (p.getSubmittedFileName() == null)
                continue;
            UUID uuid = UUID.randomUUID();
            String sfn = p.getSubmittedFileName();
            String suffix = sfn.substring(sfn.lastIndexOf("."));
            p.write("C:/photo/"+uuid.toString()+suffix);
        }
        sb.append("用户名: ").append(username).append("<br>");
        sb.append("密码: ").append(password1).append("<br>");
        sb.append("电话号码: ").append(phone).append("<br>");
        sb.append("性别: ").append(gender).append("<br>");
        sb.append("爱好: ");
        for(String h:hobby){
            sb.append(h).append("  ");
        }
        sb.append("<br>");
        resp.getWriter().println(sb);
    }
}
```

在 Register1.java 代码中,接收多上传文件的关键代码如下:

```
Collection<Part> parts = req.getParts();
for(Part p : parts){
    if (p.getSubmittedFileName() == null)
        continue;
    UUID uuid = UUID.randomUUID();
    String sfn = p.getSubmittedFileName();
    String suffix = sfn.substring(sfn.lastIndexOf("."));
    p.write("C:/photo/"+uuid.toString()+suffix);
}
```

在这段代码中,通过调用 HTTP 请求对象的 getParts 方法得到所有的 Part,然后针对每个 Part,判断是否存在上传时客户端设置的文件名,也就是如果 p.getSubmittedFileName() == null,则这个 Part 不是文件;否则利用 UUID 生成一个唯一的文件名,并将上传的文件以所生成的文件名保存在 C:/photo 目录下。

form 表单中 multipart/form-data 的本质

运行这个程序,在浏览器的地址栏输入 register1.html,然后在选择文件时选择多个文件,程序可以上传多个文件到服务端。

4.2 Servlet 下载文件到客户端

Servlet 程序可以将任何类型的文件下载到客户端程序中,客户端程序可以根据文件的 MIME 类型进行适当处理,例如,显示或保存下载的文件等。下面先从简单地下载一张图片到浏览器开始。

4.2.1 下载并显示图像

这里"下载并显示图像"中的图像存储于服务器的私有位置，例如，存储于数据库中，或者存储于服务器的某个私有目录下。客户端是不能直接访问存储于这些私有位置的图像的，必须要通过服务端程序处理后再发送给客户端程序。例如，在服务器的 C:/Temp 目录下有一些风景图像，如图 4-4 所示。

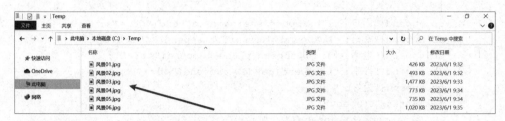

图 4-4　服务器 C:/Temp 目录下的一些风景图片

客户端浏通过服务端的 Servlet 程序随机访问其中的任何一张图像并显示。为此，在 ch04 工程中新建一个 RandomImage 程序，RandomImage.java 程序的代码如下：

```java
package com.ttt.servlet;
import jakarta.servlet.ServletException;
import jakarta.servlet.annotation.WebServlet;
import jakarta.servlet.http.HttpServlet;
import jakarta.servlet.http.HttpServletRequest;
import jakarta.servlet.http.HttpServletResponse;
import java.io.FileInputStream;
import java.io.IOException;
import java.net.URLEncoder;
import java.nio.charset.StandardCharsets;
import java.util.Random;
@WebServlet("/RandomImage")
public class RandomImage extends HttpServlet{
    @Override
    protected void doGet(HttpServletRequest req, HttpServletResponse resp)
        throws ServletException, IOException{
        Random r = new Random();
        int which = Math.abs(r.nextInt() % 5) + 1;
        String name = "风景 0" + which + ".jpg";
        String fn = "C:/Temp/" + name;
        String mime = getServletContext().getMimeType(fn);
        resp.setContentType(mime);
        try(FileInputStream fis = new FileInputStream(fn)){
            byte[] b = new byte[1024];
            while(fis.read(b)>0) {
                resp.getOutputStream().write(b);
            }
        }
    }
}
```

这个程序利用随机数发生器生成一个随机整数,并以此为基础构建要访问的图像的文件名。然后使用如下语句:

```
String mime = getServletContext().getMimeType(fn);
resp.setContentType(mime);
```

根据文件名得到文件的 MIME 类型,并设置 HTTP 响应的 MIME 类型,然后使用如下语句:

```
try(FileInputStream fis = new FileInputStream(fn)) {
    byte[] b = new byte[1024];
    while(fis.read(b)>0) {
        resp.getOutputStream().write(b);
    }
}
```

从图像源文件中读出图像数据,并写入响应对象流中,进而客户端可以接收到图像数据流并做适当处理。例如,对浏览器而言,就直接将图像显示出来。现在运行这个程序,并在浏览器中地址栏中输入 RandomImage 地址,显示如图 4-5 所示界面。

图 4-5　随机访问图像程序的运行结果

不断单击浏览器的刷新按钮,将显示不同的图像。当然,也可以将随机图像在 HTML 页面中显示出来。为此,在 ch04 程序的 html 目录下新建名为 show.html 的页面文件。show.html 的代码如下:

```
<!DOCTYPE html>
<html lang="en">
<head>
    <meta charset="UTF-8">
    <title>Title</title>
</head>
<body>
<h2 style="text-align: center">随机显示风景图像</h2>
```

```
<img src="../RandomImage" style="width: 100%; height: 100%; position: absolute"
alt=""/>
</body>
</html>
```

在 img 标签中直接引用了 RandomImage 的图像访问程序的地址,从而可在 img 标签中显示图像。运行这个程序,在浏览器地址栏输入 show.html 的地址,显示如图 4-6 所示的结果。

图 4-6　将图像嵌入 HTML 页面

不断单击浏览器的刷新按钮,将显示不同的图像。

4.2.2　下载并保存图像文件

在图 4-5 或图 4-6 的界面中,如果在图像上右击,在弹出的菜单中选择"将图像另存为"命令来保存图像,会发现保存文件的文件名是服务端程序的程序名,并且文件的后缀名也可能不正确。这是因为在 RandomImage 程序中没有正确指定文件名导致的。下面修改 RandomImage 程序,重命名新的程序为 RandomImage1。RandomImage1 设置在客户端保存文件时的文件名。RandomImage1.java 程序如下:

```
package com.ttt.servlet;
import jakarta.servlet.ServletException;
import jakarta.servlet.annotation.WebServlet;
import jakarta.servlet.http.HttpServlet;
import jakarta.servlet.http.HttpServletRequest;
import jakarta.servlet.http.HttpServletResponse;
import java.io.FileInputStream;
import java.io.IOException;
import java.net.URLEncoder;
import java.nio.charset.StandardCharsets;
```

```java
import java.util.Random;
@WebServlet("/RandomImage1")
public class RandomImage1 extends HttpServlet{
    @Override
    protected void doGet(HttpServletRequest req, HttpServletResponse resp)
            throws ServletException, IOException{
        Random r = new Random();
        int which = r.nextInt(1, 7);
        String name = "风景0" + which + ".jpg";
        String fn = "C:/Temp/" + name;
        String mime = getServletContext().getMimeType(fn);
        resp.setContentType(mime);
        resp.setHeader("content-disposition","attachment;filename="
            + URLEncoder.encode(name, StandardCharsets.UTF_8));
        try(FileInputStream fis = new FileInputStream(fn)){
            byte[] b = new byte[1024];
            while(fis.read(b)>0) {
                resp.getOutputStream().write(b);
            }
        }
    }
}
```

在代码中,通过语句 "resp.setHeader("content-disposition","attachment;filename=" + URLEncoder.encode(name, StandardCharsets.UTF_8));" 明确告知客户端用户要保存这个文件时的文件名是什么。这里,使用 UTF-8 对文件名进行编码,以解决文件名中存在中文字符时的乱码问题。在运行这个程序之前,先修改 show.html 代码,将其 img 标签的 src 属性指向 RandomImage1:

```html
<img src="../RandomImage1" style="width: 100%; height: 100%; position: absolute" alt=""/>
```

运行这个程序,右击图像并保存图像到文件时,会发现文件名正是程序中指定的文件名和后缀名。

4.2.3 下载和保存任意类型的文件

与下载显示图像或者下载保存图像类似的方式,可以下载保存任意类型的文件。下载保存文件的关键点在于:其一,必须使用 setContentType() 方法正确设置文件的 MIME 类型;其二,使用 setHeader() 方法正确设置文件的文件名。下面举个例子说明如何下载和保存任意类型的文件。为此,在 ch04 工程中新建一个名为 FileDownloader 的服务端程序,FileDownloader.java 程序的代码如下:

```java
package com.ttt.servlet;
import jakarta.servlet.ServletException;
import jakarta.servlet.annotation.WebServlet;
import jakarta.servlet.http.HttpServlet;
import jakarta.servlet.http.HttpServletRequest;
import jakarta.servlet.http.HttpServletResponse;
```

```java
import java.io.FileInputStream;
import java.io.FileNotFoundException;
import java.io.IOException;
import java.net.URLEncoder;
import java.nio.charset.StandardCharsets;
@WebServlet("/FileDownloader")
public class FileDownloader extends HttpServlet{
    @Override
    protected void doGet(HttpServletRequest req, HttpServletResponse resp)
        throws ServletException, IOException{
        req.setCharacterEncoding("UTF-8");
        String which = req.getParameter("which");
        if ((which == null) || (which.isBlank())) {
            resp.setContentType("text/html; charset=utf-8");
            resp.sendError(500, "缺少必要的参数数据");
            return;
        }
        //这里是 Java 17 才有的功能,需要在 pom.xml 中将 11 改成 17
        String name = switch (which) {
            case "PDF" -> "示例 Word 文档.pdf";
            case "WORD" -> "示例 Word 文档.docx";
            case "IMAGE1" -> "风景 01.jpg";
            case "IMAGE2" -> "风景 02.jpg";
            case "IMAGE3" -> "风景 03.jpg";
            default -> "wheat.jpg";
        };
        String fn = "C:/Temp/" + name;
        try(FileInputStream fis = new FileInputStream(fn)){
            String mime = getServletContext().getMimeType(fn);
            resp.setContentType(mime);
            resp.setHeader("content-disposition","attachment;filename="
                + URLEncoder.encode(name, StandardCharsets.UTF_8));
            byte[] b = new byte[1024];
            while(fis.read(b)>0) {
                resp.getOutputStream().write(b);
            }
        }
        catch (FileNotFoundException e){
            resp.setContentType("text/html; charset=utf-8");
            resp.sendError(500, e.toString());
        }
    }
}
```

在访问这个 Servlet 程序时需要给出一个名称为 which 的参数,用以指明要下载的文件。可用的参数值包括:①PDF,用于下载 PDF 文件;②WORD,用于下载 Word 文档文件;③IMAGE1,用于下载编号为 1 的风景图像;④IMAGE2,用于下载编号为 2 的风景图像;⑤IMAGE3,用于下载编号为 3 的风景图像。程序根据参数 which 的值,打开相应的文件并发送到客户端。程序同时也对可能出现的异常情况做了处理。

现在编写客户端 HTML 程序 downloader.html,它给出可下载列表的同时,也直接使用 img 标签访问 FileDownloader 程序并显示一张图像。downloader.html 程序代码如下：

```html
<!DOCTYPE html>
<html lang="en">
<head>
    <meta charset="UTF-8">
    <title>文件下载和图像显示示例</title>
</head>
<body>
<h2 style="text-align: center"><a href="../FileDownloader?which=PDF">下载 PDF 文件</a></h2>
<h3 style="text-align: center">
    <a href="../FileDownloader?which=WORD">下载 WORD 文件</a>
    <a href="../FileDownloader?which=IMAGE1">下载图像文件 1</a>
    <a href="../FileDownloader?which=IMAGE2">下载图像文件 2</a>
    <a href="../FileDownloader?which=IMAGE3">下载图像文件 3</a>
</h3>
<img src="../FileDownloader?which=IMAGE1" alt=""/>
</body>
</html>
```

在这个 HTML 页面中，使用<a>标签指向 FileDownloader 并用于下载文件，同时，也使用了<img>标签指向 FileDownloader 并用于直接显示图像。运行这个程序，在浏览器地址栏输入 downloader.html，显示如图 4-7 所示的结果。

图 4-7　下载任何类型的文件运行结果

4.3　案例：美图分享

生活中拍摄的美景照片可以分享给其他人：清晨美丽的朝霞、暮色美丽的晚霞、雄伟俊美的山川、清澈流动的小溪还有一群群叽叽喳喳的小鸟……

4.3.1 案例目标

编写一个 Java Web 程序实现风景图像分享。用户可以将拍摄的风景照片上传到这个系统,任何人都可以查看和下载别人上传的美丽风景图像。

4.3.2 案例分析

这个程序需要具备两个主要功能:其一,用户可以将拍摄的风景图像上传到系统中;其二,用户可以查看或者下载上传的风景图像。因此,在程序的首页面上,需要显示两个功能链接:上传风景图像和浏览/下载风景图像。可以考虑使用如图 4-8 所示的首页面布局。

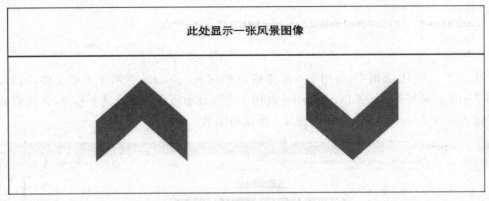

图 4-8　美图分享首页面布局

第一个图标表示上传风景图像,第二个图标表示浏览/下载风景图像。当单击第一个图标时,显示风景图像分享页面,如图 4-9 所示。

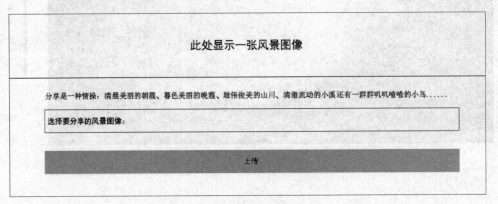

图 4-9　风景图像上传分享页面

在这个界面中,用户可以选择一张或多张图像文件并上传到服务器上。当单击第二个图标时,显示风景图像列表页面,如图 4-10 所示。

在这个界面中,用户单击任何一张图像,将在浏览器中显示所单击图像的原始图像,同时,用户可以右击图像,从而保存图像文件到本地。基于这样的分析结果,程序包括如下文件。

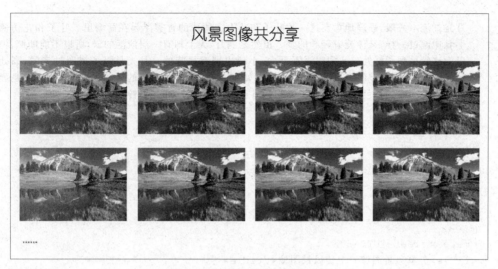

图 4-10 风景图像分享列表页面

(1) sindex.html:程序首页面。
(2) uploader.html:图像上传页面。
(3) share.css:页面相关布局文件。
(4) Receiver.java:接收客户端上传的图像,并将它们保存在服务器指定的位置。
(5) Share.java:将服务器中的图像以列表的形式在客户端显示出来。

4.3.3 案例实施

先准备好页面上需要的图标文件,并将它们放置在 images 子目录下,再编写 sindex.html 首页面程序。为此,在 ch04 工程的 html 子目录下创建这个文件。完成后的 sindex.html 代码如下:

```html
<!DOCTYPE html>
<html lang="en">
<head>
  <meta charset="UTF-8">
  <title>风景图像分享</title>
  <link href="../css/share.css" rel="stylesheet" type="text/css" />
</head>
<body>
  <img class="banner" src="../images/banner.jpg" alt=""/>
  <div>
    <a href="uploader.html">
      <img class="image" src="../images/upload.png" title="单击分享风景图像" alt="" />
    </a>
    <a href="../Share">
      <img class="image" src="../images/download.png" title="单击浏览风景图像" alt="" />
    </a>
  </div>
```

```html
    <p>
        月光如流水一般，静静地泻在这一片叶子和花上。薄薄的青雾浮起在荷塘里。叶子和花仿佛在牛乳中洗过一样；又像笼着轻纱的梦。虽然是满月，天上却有一层淡淡的云，所以不能朗照；但我以为这恰是到了好处——酣眠固不可少，小睡也别有风味的。月光是隔了树照过来的，高处丛生的灌木，落下参差的斑驳的黑影，峭楞楞如鬼一般；弯弯的杨柳的稀疏的倩影，却又像是画在荷叶上。塘中的月色并不均匀；但光与影有着和谐的旋律，如梵婀铃上奏着的名曲。
    </p>
</body>
</html>
```

uploader.html 页面代码如下：

```html
<!DOCTYPE html>
<html lang="en">
<head>
    <meta charset="UTF-8">
    <title>分享风景图像—上传风景图像</title>
    <link href="../css/share.css" rel="stylesheet" type="text/css" />
</head>
<body>
    <img class="banner" src="../images/banner.jpg" alt=""/>
    <p class="ppp">
        分享是一种情操：清晨美丽的朝霞、暮色美丽的晚霞、雄伟俊美的山川、清澈流动的小溪还有一群群叽叽喳喳的小鸟……
    </p>
    <div>
        <form enctype="multipart/form-data" method="POST" action="../Receiver">
            <input type="file" multiple id="scene" name="scene" placeholder="选择...">
            <input type="submit" value="提交">
        </form>
    </div>
</body>
</html>
```

以上两个页面文件中，用到了 share.css 的布局文件。share.css 代码如下：

```css
.banner{
    width: 100%;
    height: 200px;
}
div{
    width: 100%;
    text-align: center
}
.image{
    width: 200px;
    height: 200px;
    margin: 30px 150px;
}
.image:hover {
    transform: translateY(-10px);
```

```css
        filter: drop-shadow(.5rem .5rem 1rem);
        background-color: #45a049;
    }
    .ppp{
        margin: 80px 30px 30px 10px;
    }
    input[type=file]{
        width: 100%;
        padding: 12px 20px;
        margin: 8px 0;
        display: inline-block;
        border: 1px solid #ccc;
        border-radius: 4px;
        box-sizing: border-box;
    }
    input[type=submit]{
        width: 100%;
        background-color: #4CAF50;
        color: white;
        padding: 14px 20px;
        margin: 8px 0;
        border: none;
        border-radius: 4px;
        cursor: pointer;
    }
    input[type=submit]:hover{
        background-color: #45a049;
    }
```

Receiver.java 程序用于接收从 uploader.html 页面发送的图像文件,Receiver.java 代码如下:

```java
package com.ttt.servlet;
import jakarta.servlet.ServletException;
import jakarta.servlet.annotation.MultipartConfig;
import jakarta.servlet.annotation.WebServlet;
import jakarta.servlet.http.HttpServlet;
import jakarta.servlet.http.HttpServletRequest;
import jakarta.servlet.http.HttpServletResponse;
import jakarta.servlet.http.Part;
import java.io.IOException;
import java.util.Collection;
import java.util.UUID;
@MultipartConfig
@WebServlet("/Receiver")
public class Receiver extends HttpServlet{
    @Override
    protected void doPost(HttpServletRequest req, HttpServletResponse resp)
      throws ServletException, IOException{
        req.setCharacterEncoding("utf-8");
```

```java
            Collection<Part> parts = req.getParts();
            for(Part p : parts){
                if (p.getSubmittedFileName() == null)
                    continue;
                UUID uuid = UUID.randomUUID();
                String sfn = p.getSubmittedFileName();
                String suffix = sfn.substring(sfn.lastIndexOf("."));
                p.write("C:/SceneShare/"+uuid.toString()+suffix);
            }
            resp.setContentType("text/html; charset=utf-8");
            resp.getWriter().println("<img src='images/success.png' alt='' " +
                    "style='clear: both; display: block; margin: auto;'/>");
    }
}
```

Share.java 程序以列表形式显示已经上传的图像文件。Share.java 代码如下:

```java
package com.ttt.servlet;
import jakarta.servlet.ServletException;
import jakarta.servlet.annotation.WebServlet;
import jakarta.servlet.http.HttpServlet;
import jakarta.servlet.http.HttpServletRequest;
import jakarta.servlet.http.HttpServletResponse;
import java.io.File;
import java.io.FileInputStream;
import java.io.IOException;
import java.net.URLEncoder;
import java.nio.charset.StandardCharsets;
@WebServlet("/Share")
public class Share extends HttpServlet{
    @Override
    protected void doGet(HttpServletRequest req, HttpServletResponse resp)
        throws ServletException, IOException{
        req.setCharacterEncoding("utf-8");
        String name = req.getParameter("name");
        if ((name != null) && (!name.isBlank())) {
            String fn = "C:/SceneShare/" + name;
            try(FileInputStream fis = new FileInputStream(fn)){
                String mime = getServletContext().getMimeType(fn);
                resp.setContentType(mime);
                resp.setHeader("content-disposition","attachment;filename="+
                        URLEncoder.encode(name, StandardCharsets.UTF_8));
                byte[] b = new byte[1024];
                while(fis.read(b)>0) {
                    resp.getOutputStream().write(b);
                }
            }
            return;
        }
        File dir = new File("C:/SceneShare/");
        String[] ifn = dir.list();
        if ((ifn == null) || (ifn.length == 0)) return;
```

```
resp.setContentType("text/html; charset=utf-8");
resp.getWriter().println("<h2 style=\"text-align: center\">风景图像共分
享</h2>");
for(String n : ifn) {
    String s ="./Share?name=" + n;
    resp.getWriter().println(
        "<a href=\"" + s + "\">" +
        "<img src=\"" + s + "\" alt=\"\" " +
            "style=\"margin: 5px; width: 23%; height: 30%; float: 
            left\"/>" + " </a>");
    }
  }
}
```

观察 Share.java 程序,发现程序里有太多 HTML 标签代码,使得程序的可读性非常不好。在后续章节中,会采用新的技术解决这个问题。现在运行这个程序,在浏览器的地址栏输入 sindex.html,将显示如图 4-11 所示的结果。

单击相应的功能,即可完成风景图像的分享和查看/下载。

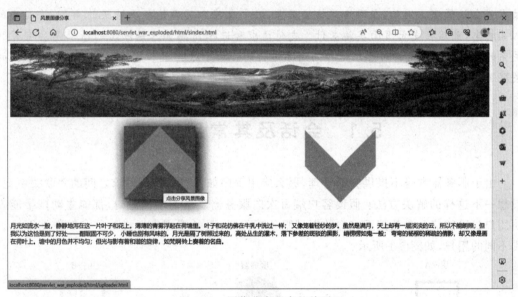

图 4-11　图像分享程序的首页面

4.4　练习:完善书籍录入程序

完善书籍信息录入程序。要求首先制作一个书籍录入页面,该页面除了可以录入书籍的基本信息,包括书籍名称、出版社、编者、价格、书籍介绍等基本信息以外,还要求录入如下新的信息:书籍的封面、书籍的类型。当用户在页面上录入信息并单击"提交"按钮后,将书籍信息保存在后台服务器中。书籍信息保存到服务器后,返回一个页面,在该页面中,用美观的方式显示书籍的完整信息。

第 5 章 会 话 管 理

HTTP 是无状态协议。"无状态"的含义是：假设客户端向服务器发送了两个 HTTP 请求，暂且分别称为请求 A 和请求 B，服务器不会也不能从这两个请求信息本身建立这两个请求之间的关联关系。简单来说就是，对服务器而言，请求 A 和请求 B 是完全没有任何关系的：当服务器收到请求 A 时，它会根据请求 A 的参数和要求完成必要的操作，然后将处理结果发送到客户端；当服务器收到请求 B 时，它会根据请求 B 的参数和要求完成必要的操作，然后将处理结果发送到客户端。

然而在实际应用中，请求之间是存在关联关系的，称这种对请求的关联关系进行管理的技术为会话管理。例如，在网络购物应用中，用户在某个页面选择要购买的商品，将商品加入购物车，最后在支付页面实施支付。支付请求和选择商品加入购物车的请求是具有关联关系的：支付请求是针对已经选择的商品进行的。由于服务器本身不提供不同请求之间的关联关系管理，这就要求应用程序自己对会话进行管理。

5.1 会话及其常用技术

由于服务器本身不提供会话管理，那么应用程序如何自行管理请求之间的关联关系呢？设想一下这样的解决方法：假设客户端每次向服务器发送请求消息时，都能主动携带能表明用户身份的信息参数甚至完整数据到服务端程序，服务端程序通过这些参数数据即可识别不同的用户，如图 5-1 所示。

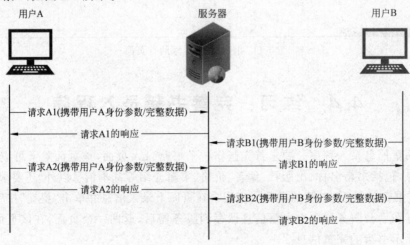

图 5-1 会话管理技术

从图 5-1 可以看出，如果每个用户向服务器发送请求时，每次都主动携带能表明自己身份参数乃至完整的数据信息到服务器，那么，服务端程序即可识别不同的用户，进而可以解决会话管理问题。在这样的指导思想下，出现了两种常用的具体解决方法。其一，Cookie 技术：每个请求均携带用户的完整数据信息；其二，Session 技术：每个请求均携带表明用户身份的参数信息。

5.2 Cookie 技 术

Cookie 是一种常用会话管理技术，在这种技术中，客户端向服务端发送请求信息时，每次均携带完整的用户数据信息。

5.2.1 什么是 Cookie

Cookie 翻译成中文是小饼干。在计算机领域术语中，它代表的是保存在客户端的一小段数据：是服务端程序通过 HTTP 响应头发送给客户端的一小段数据，根据 HTTP 约定要求，在客户端再次访问同一个服务器时，浏览器需要再次将这一小段数据传回到服务端程序。基于这样的约定要求，可以使用 Cookie 进行会话管理。例如，在购物应用中，对用户在购物页面选择的商品，可以通过 Cookie 将用户所选商品信息发送到客户端，当用户进行支付结算时，由于客户端会将用户所选商品发送到服务端程序，所以，服务端程序可以基于这些信息完成用户的结算操作。Java Web 对 Cookie 技术提供了很好的支持。

5.2.2 Cookie 类

为了支持应用程序使用 Cookie 技术实现会话管理，Java Web 提供了 Cookie 类。Cookie 具有名字和值，还可以具有如下属性：最大有效期、版本、描述、URL 路径和域名限定符等。Cookie 类的常用方法及其含义如表 5-1 所示。

表 5-1 Cookie 类的常用方法及其含义

序号	方 法 名	描 述
1	Cookie(String name, String value)	Cookie 构造函数，Cookie 具有名字和值，还可以通过下列方法设置一系列属性
2	void setDomain(String domain)	设置 Cookie 的域名。默认情况下，Cookie 的域名就是创建 Cookie 的服务器域名
3	void setMaxAge(int expiry)	设置 Cookie 的最大有效时长。单位是秒。正值表示 Cookie 的最大有效时长；负值表示当浏览器退出时，立刻删除这个 Cookie；0 表示删除这个 Cookie
4	void setPath(String uri)	设置 Cookie 的有效 URL 路径。所谓有效 URL 路径，是指当浏览器访问这个 URL 时，浏览器才会把 Cookie 发送到服务端
5	void setValue(String newValue)	设置 Cookie 的值

续表

序号	方法名	描述
6	void setVersion(int v)	设置 Cookie 的版本
7	String getDomain()	返回 Cookie 的域名
8	int getMaxAge()	返回 Cookie 的最大有效时长
9	String getName()	返回 Cookie 的名字
10	String getPath()	返回浏览器发送这个 Cookie 到服务器的 URL 路径
11	String getValue()	返回 Cookie 的值
12	int getVersion()	返回 Cookie 的版本

在客户端能够发送 Cookie 到服务端之前,服务端程序需要在 HttpServletResponse 对象中使用 addCookie()方法将 Cookie 添加到 HttpServletResponse 的响应头中。客户端收到这个 HTTP 响应后,会按照 Cookie 的最大有效时长将 Cookie 保存到客户端,并且当再次向服务端发送 HTTP 请求时,按约定将 Cookie 发送到服务端程序。服务端程序通过 HttpServletRequest 对象的 getCookies()方法获取 Cookie 的值并进行相应处理。

5.2.3 使用 Cookie 实现会话管理举例

举一个例子说明 Cookie 的使用。这个程序包括今日任务录入页面和今日任务显示页面:用户在录入页面录入相关信息,单击"提交"按钮后,将所填写信息发送到服务端程序,服务器程序创建 Cookie 并将信息通过 Cookie 保存在客户端,然后显示一个查看信息页面,单击这个页面的"查看今日任务"链接,可以查看已经录入的今日任务信息。为此,再新建 ch05 Java Web 工程。创建完成的 ch05 工程如图 5-2 所示。

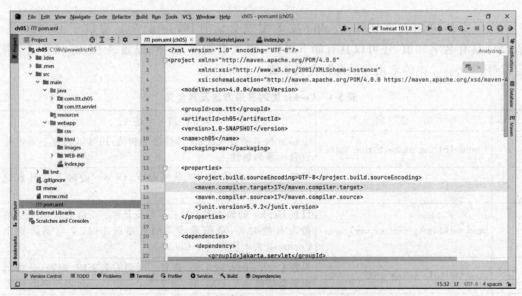

图 5-2 新建的 ch05 Java Web 工程

该程序包括如下几个文件。

(1) schedule.html：录入进入任务的页面文件。

(2) style.css：页面布局文件。

(3) SaveInCookie.java：Servlet 程序。接收 schedule.html 页面录入的信息，创建 Cookie 对象，并将 Cookie 发送到客户端浏览器保存。

(4) ViewCookie.java：Servlet 程序。查看从客户端浏览器发送到服务端的 Cookie 数据。

schedule.html 页面的代码如下：

```html
<!DOCTYPE html>
<html lang="en">
<head>
  <meta charset="UTF-8">
  <title>录入今日任务</title>
  <link href="../css/style.css" rel="stylesheet" type="text/css" />
</head>
<body>
  <h2 style="text-align: center">请录入今日任务</h2>
  <div>
    <form enctype="application/x-www-form-urlencoded"
          method="POST" action="../SaveInCookie">
      <label for="title">任务名称</label>
      <input class="title" type="text" id="title" name="title" placeholder="任务名称...">
      <label for="desc">任务具体内容
        <textarea class="title desc" id="desc" name="desc" rows="5" cols="30"
                  wrap="soft" placeholder="任务描述..."></textarea>
      </label>
      <input type="submit" value="提交">
    </form>
  </div>
</body>
</html>
```

这个页面代码中包括一个 form，在 form 表单中的输入标签中可以录入任务名称和任务内容相关信息，然后提交给 SaveInCookie 服务端代码处理。这个页面用到的布局文件 style.css 内容如下：

```css
.title{
    width: 100%;
    padding: 12px 20px;
    margin: 8px 0;
    display: inline-block;
    border: 1px solid #ccc;
    border-radius: 4px;
    box-sizing: border-box;
}
.desc{
    height: 300px;
```

```css
}
input[type=submit]{
    width: 100%;
    background-color: #4CAF50;
    color: white;
    padding: 14px 20px;
    margin: 8px 0;
    border: none;
    border-radius: 4px;
    cursor: pointer;
}
input[type=submit]:hover{
    background-color: #45a049;
}
div{
    border-radius: 5px;
    background-color: #f2f2f2;
    padding: 20px;
}
```

SaveInCookie.java 服务端程序如下：

```java
package com.ttt.servlet;
import jakarta.servlet.ServletException;
import jakarta.servlet.annotation.WebServlet;
import jakarta.servlet.http.Cookie;
import jakarta.servlet.http.HttpServlet;
import jakarta.servlet.http.HttpServletRequest;
import jakarta.servlet.http.HttpServletResponse;
import java.io.IOException;
import java.net.URLEncoder;
import java.nio.charset.StandardCharsets;
@WebServlet("/SaveInCookie")
public class SaveInCookie extends HttpServlet{
    @Override
    protected void doPost(HttpServletRequest req, HttpServletResponse resp)
      throws ServletException, IOException{
        req.setCharacterEncoding("UTF-8");
        String title = req.getParameter("title");
        String desc = req.getParameter("desc");
        if ((title == null) || (title.isBlank()) || (desc == null) || desc.isBlank())
        {
            resp.setContentType("text/html; charset=utf-8");
            resp.getWriter().println("任务名称及任务描述都不能为空");
            return;
        }
        Cookie cookie_title = new Cookie("title",
           URLEncoder.encode(title, StandardCharsets.UTF_8));
        cookie_title.setMaxAge(24 * 60 * 60);
        Cookie cookie_desc = new Cookie("desc",
```

```
        URLEncoder.encode(desc, StandardCharsets.UTF_8));
    cookie_desc.setMaxAge(24 * 60 * 60);
    resp.addCookie(cookie_title);
    resp.addCookie(cookie_desc);
    resp.setContentType("text/html; charset=utf-8");
    resp.getWriter().println("<h2 style='text-align: center'>" +
        "<a href='./ViewCookie'>查看今日任务</a></h2>");
    }
}
```

在这个程序中,首先获取从客户端发送的参数数据,然后使用如下语句:

```
Cookie cookie_title = new Cookie("title",
    URLEncoder.encode(title, StandardCharsets.UTF_8));
cookie_title.setMaxAge(24 * 60 * 60);
```

创建名字为 title 的 Cookie 对象,并设置其最长有效时长为一天。类似地,创建类名字为 desc 的 Cookie 对象。将这两个 Cookie 添加到 HttpServletResponse 对象中,进而通过 HTTP 响应将 Cookie 发送到客户端保存起来。最后,使用如下语句:

```
resp.getWriter().println("<h2 style='text-align: center'>" +
    "<a href='./ViewCookie'>查看今日任务</a></h2>");
```

在客户端浏览器显示一个链接,单击这个链接将访问 ViewCookie,显示 Cookie 数据。ViewCookie.java 代码如下:

```
package com.ttt.servlet;
import jakarta.servlet.ServletException;
import jakarta.servlet.annotation.WebServlet;
import jakarta.servlet.http.Cookie;
import jakarta.servlet.http.HttpServlet;
import jakarta.servlet.http.HttpServletRequest;
import jakarta.servlet.http.HttpServletResponse;
import java.io.IOException;
import java.net.URLDecoder;
import java.nio.charset.StandardCharsets;
import java.util.Arrays;
@WebServlet("/ViewCookie")
public class ViewCookie extends HttpServlet{
    @Override
    protected void doGet(HttpServletRequest req, HttpServletResponse resp)
        throws ServletException, IOException{
        req.setCharacterEncoding("UTF-8");
        Cookie[] cookies = req.getCookies();
        String title = "", desc = "";
        for(Cookie c : cookies){
            if (c.getName().equalsIgnoreCase("title")) {
                title = c.getValue();
            }
            else if (c.getName().equalsIgnoreCase("desc")) {
                desc = c.getValue();
```

```
        }
    }
    resp.setContentType("text/html; charset=utf-8");
    String head = """
            <!DOCTYPE html>
              <html lang="en">
                <head>
                  <meta charset="UTF-8">
                  <title>录入今日任务</title>
                  <link href="./css/style.css" rel="stylesheet"
                  type="text/css" />
                </head>
                <body>
            """;
    resp.getWriter().println(head);
    resp.getWriter().println("<h2 style=\"text-align: center\">今日任务
    </h2>");
    resp.getWriter().println("<div>");
    resp.getWriter().println("<input class='title' type='text' value='" +
            URLDecoder.decode(title, StandardCharsets.UTF_8) +
            "' contenteditable='false'>");
    resp.getWriter().println("<textarea class='title desc' contenteditable=
    'false'>" + URLDecoder.decode(desc, StandardCharsets.UTF_8));
    resp.getWriter().println("</textarea></div></body></html>");
    }
}
```

这个程序首先使用如下语句:

```
for(Cookie c : cookies) {
    if (c.getName().equalsIgnoreCase("title")) {
        title = c.getValue();
    }
    else if (c.getName().equalsIgnoreCase("desc")) {
        desc = c.getValue();
    }
}
```

从HTTP的请求中获取Cookie对象。然后使用如下语句:

```
resp.getWriter().println("<input class='title' type='text' value=
'" + URLDecoder.decode(title, StandardCharsets.UTF_8) + "'
contenteditable='false'>");
resp.getWriter().println("<textarea class='title desc' contenteditable='false'>" +
    URLDecoder.decode(desc, StandardCharsets.UTF_8));
```

将title和desc的值显示在客户端浏览器页面中。现在运行这个程序,在浏览器地址栏输入schedule.html,显示如图5-3所示的结果。

在图5-3的界面中录入相关信息,然后单击"提交"按钮,显示如图5-4所示的界面。

在图5-4的界面中,单击链接即可显示今日任务数据,如图5-5所示。

图 5-3　今日任务首页面

图 5-4　显示查看数据链接页面

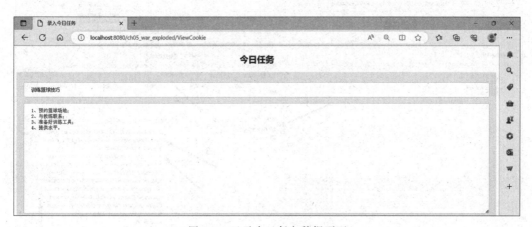

图 5-5　显示今日任务数据页面

5.2.4　Cookie 观察和 Cookie 使用注意事项

观察一下客户端与服务器之间采用 Cookie 进行会话管理的过程。在图 5-3 的界面中，打开浏览器的"开发人员工具"，然后，单击"提交"按钮，显示如图 5-6 所示的界面。

从 5-6 可以看出，在服务端程序 SaveInCookie 代码中创建的 Cookie 对象已经通过 HTTP 响应发送到了浏览器中：这个响应头的名字是 Set-Cookie。再单击"查看今日任务"链接，此时，浏览器会将这两个 Cookie 通

Cookie 观察

过HTTP请求送回到服务器中,如图5-7所示。

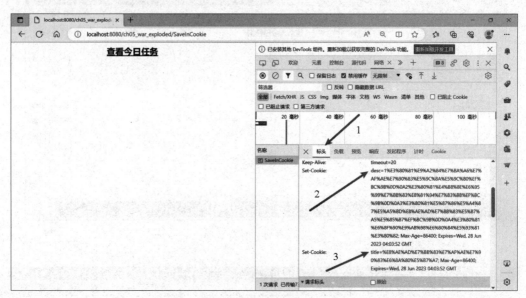

图5-6　服务端发送到浏览器的Cookie

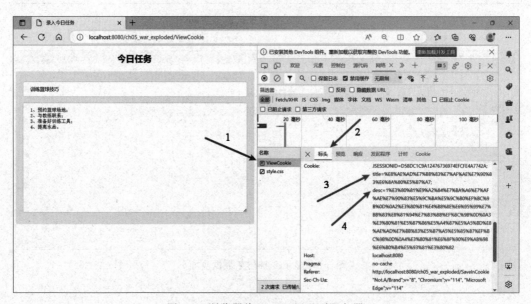

图5-7　浏览器将Cookie回送到服务器

从图5-7可以看出,当浏览器向服务器发送Cookie时,请求头的名字是Cookie。

由于Cookie数据保存在客户端,所以,在Cookie中不应存储涉及安全相关的数据,如密码之类的数据。另外,针对每个Web服务器,浏览器只能保存20个Cookie数据,并且,浏览器能够保存的Cookie的总数不能超过300个,且Cookie数据总量不能超过4KB。同时,针对浏览器而言,用户可以关闭Cookie数据,也就是不允许在浏览器中使用Cookie。因此,在编程实践中,要慎用Cookie跟踪和管理会话。为此,需要更好的技术来管理会话。这个技术就是Session。

5.3 Session 技术

Session 是最常用的会话跟踪管理技术。与 Cookie 将所有数据保存在客户端不同，Session 技术不会将数据保存在客户端。服务端程序一旦启用 Session 技术来维护会话，服务端程序会生成一个唯一的称为 session-id 的编号，并将这个编号通过 HTTP 响应头发送到客户端。客户端在后续的每次请求中，都会自动携带该 session-id 编号到服务端程序，服务端程序以此为依据来识别不同的用户。不仅如此，服务端程序还可以以此 session-id 编号为基础，在服务端程序中保存与该 session-id 对应的用户的所有数据。因此，使用 Session 技术跟踪和管理会话有如下优点：其一，客户端和服务端程序之间的交互数据量少；其二，使用 Session 技术跟踪和管理会话更为安全。

5.3.1 HttpSession 接口

Java Web 为使用 Session 技术实现会话跟踪和管理提供了 HttpSession 接口。服务端 Servlet 程序可以调用 HttpServletRequest 接口的 getSession() 方法获得一个 HTTP Session 接口对象。HttpServletRequest 接口提供了两种获取 HttpSession 接口的方法。

(1) HttpSession getSession(boolean create)：检查是否存在与 HttpServletRequest 接口对象关联的 HttpSession 接口对象。如果存在，则返回这个已经存在的 HttpSession 接口对象；如果不存在并且参数 create 为 true，则创建一个新的 HttpSession 对象并返回，否则返回 null。

(2) HttpSession getSession()：等价于 getSession(true)。

一旦获得了 HttpSession 接口对象，服务端程序即可方便地对会话进行访问和管理。HttpSession 接口的常用方法如表 5-2 所示。

表 5-2 HttpSession 接口的常用方法

序号	方法名	描述
1	Object getAttribute(String name)	返回存储在 Session 中指定 name 的属性值。如果 Session 中不存在指定 name 的属性，则返回 null
2	Enumeration<String> getAttributeNames()	返回在 Session 中存储的书友属性的 name
3	long getCreationTime()	返回 Session 被创建的时间，以自 1970 年 1 月 1 日零时起的毫秒数计算
4	String getId()	返回 Session 的 ID 号码
5	long getLastAccessedTime()	返回 Session 最近一次被客户端发送到服务端的时间，以自 1970 年 1 月 1 日零时起的毫秒数计算
6	int getMaxInactiveInterval()	返回 Session 不活动的最长时间以秒为单位，当 Session 超过这个时间未被使用时，服务区将清除这个 Session
7	ServletContext getServletContext()	返回与 Session 关联的应用程序的上下文

续表

序号	方法名	描述
8	void invalidate()	释放 Session 使之失效
9	boolean isNew()	如果 Session 是新创建的则返回 true,否则返回 false
10	void removeAttribute(String name)	从 Session 中移除指定的 name 属性
11	void setAttribute(String name, Object value)	在 Session 中设置名字为 name 的属性的值为 value
12	void setMaxInactiveInterval(int interval)	设置 Session 的最长不活动时间,单位为秒

使用 HttpSession 管理和维护会话的一般过程是:服务端程序通过调用 HttpServletRequest 接口对象的 getSession() 方法获得一个 HttpSession 接口对象,然后在 HttpSession 对象中使用 setAttribute() 方法保存需要跨请求可使用的数据,这些数据一旦保存在 HttpSession 对象中,对同一个用户的请求,均可以在获得 HttpSession 对象后再次使用其 getAttribute() 方法访问已经保存在 HttpSession 中的数据。

5.3.2 使用 HttpSession 管理会话举例

举一个例子说明如何使用 Session 来管理和维护会话。这个例子与 5.2.3 小节的例子完成同样的功能,也就是说,实现今日任务的录入和今日任务的显示,但是,这里使用 Session 来管理和维护会话:将录入的今日任务信息保存到 Session 中。因为今日任务信息已经保存在 Session 中,因此,这些信息是可以跨请求访问的。该程序包括如下几个文件。

(1) schedule2.html:录入任务的页面文件。
(2) style.css:页面布局文件。
(3) SaveInSession.java:Servlet 程序。接收 schedule2.html 页面录入的信息,创建 HttpSession 接口对象,并把今日任务信息保存到 Session 中。
(4) ViewSession.java:Servlet 程序。查看保存在 Session 中的今日任务信息。

schedule2.html 页面代码与 5.2.3 小节的 schedule.html 类似,只是将 schedule.html 代码中的:

```
<form enctype="application/x-www-form-urlencoded"
    method="POST" action="../SaveInCookie">
```

修改为

```
<form enctype="application/x-www-form-urlencoded"
    method="POST" action="../SaveInSession">
```

即可,style.css 样式代码保持不变。SaveInSession.java 的代码如下:

```
package com.ttt.servlet;
import jakarta.servlet.ServletException;
import jakarta.servlet.annotation.WebServlet;
import jakarta.servlet.http.HttpServlet;
import jakarta.servlet.http.HttpServletRequest;
import jakarta.servlet.http.HttpServletResponse;
import jakarta.servlet.http.HttpSession;
```

```java
import java.io.IOException;
@WebServlet("/SaveInSession")
public class SaveInSession extends HttpServlet{
    @Override
    protected void doPost(HttpServletRequest req, HttpServletResponse resp)
        throws ServletException, IOException{
        req.setCharacterEncoding("UTF-8");
        String title = req.getParameter("title");
        String desc = req.getParameter("desc");
        if ((title == null) || (title.isBlank()) || (desc == null) || desc.isBlank())
        {
            resp.setContentType("text/html; charset=utf-8");
            resp.getWriter().println("任务名称及任务描述都不能为空");
            return;
        }
        HttpSession httpSession = req.getSession(true);
        httpSession.setMaxInactiveInterval(60 * 60);
        httpSession.setAttribute("title", title);
        httpSession.setAttribute("desc", desc);
        resp.setContentType("text/html; charset=utf-8");
        resp.getWriter().println("<h2 style='text-align: center'>" +
            "<a href='./ViewSession'>查看今日任务</a></h2>");
    }
}
```

在这段代码中,使用如下语句:

```java
HttpSession httpSession = req.getSession(true);
httpSession.setMaxInactiveInterval(60 * 60);
```

获得 HttpSession 接口对象,并设置 Session 的最大不活动时长为 1 小时。然后使用如下语句:

```java
httpSession.setAttribute("title", title);
httpSession.setAttribute("desc", desc);
```

将今日任务的 title 和 desc 信息保存到 Session 中。最后返回一个能够查看今日任务的页面。当用户单击"查看今日任务"链接时,将访问 ViewSession 服务端程序。ViewSession.java 代码如下:

```java
package com.ttt.servlet;
import jakarta.servlet.ServletException;
import jakarta.servlet.annotation.WebServlet;
import jakarta.servlet.http.*;
import java.io.IOException;
import java.net.URLDecoder;
import java.nio.charset.StandardCharsets;
import java.util.Arrays;
@WebServlet("/ViewSession")
public class ViewSession extends HttpServlet{
    @Override
    protected void doGet(HttpServletRequest req, HttpServletResponse resp)
```

```
            throws ServletException, IOException{
              req.setCharacterEncoding("UTF-8");
              HttpSession httpSession = req.getSession();
              resp.setContentType("text/html; charset=utf-8");
              String head= """
                  <!DOCTYPE html>
                    <html lang="en">
                    <head>
                      <meta charset="UTF-8">
                      <title>显示今日任务</title>
                      <link href="./css/style.css" rel="stylesheet"
                      type="text/css" />
                    </head>
                    <body>
                  """;
              resp.getWriter().println(head);
              resp.getWriter().println("<h2 style=\"text-align: center\">今日任务
              </h2>");
              resp.getWriter().println("<div>");
              resp.getWriter().println("<input class='title' type='text' value=
                '" + httpSession.getAttribute("title") + "'
                contenteditable='false'>");
              resp.getWriter().println("<textarea class='title desc' contenteditable=
              'false'>" +
                  httpSession.getAttribute("desc"));
              resp.getWriter().println("</textarea></div></body></html>");
          }
      }
```

在这个程序中，使用如下语句：

```
HttpSession httpSession = req.getSession();
```

从 HTTP 请求中获得 HttpSession 接口对象。然后使用如下语句：

```
resp.getWriter().println("<input class='title' type='text' value='" +
    httpSession.getAttribute("title") + "' contenteditable='false'>");
resp.getWriter().println("<textarea class='title desc' contenteditable=
'false'>" + httpSession.getAttribute("desc"));
```

从 HttpSession 中获取保存的 title 属性和 desc 属性的值，并显示在浏览器页面中。运行这个程序，在浏览器地址栏输入 schedule2.html，将得到与 5.2.3 小节一致的运行效果。

5.3.3　Session 观察

Session 观察

现在观察一下客户端与服务端基于 Session 技术的交互过程。在浏览器地址栏输入 schedule2.html，打开浏览器的"开发人员工具"，并在相应的输入框中输入相关信息，显示如图 5-8 所示的结果。

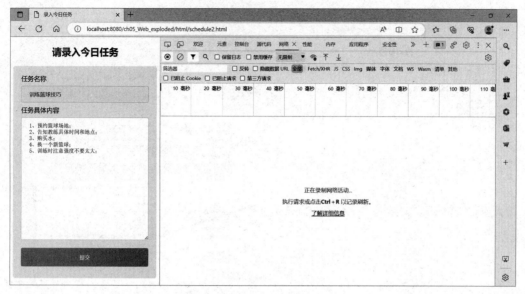

图 5-8　录入今日任务首页面

在图 5-8 的页面中单击"提交"按钮,显示如图 5-9 所示的界面。

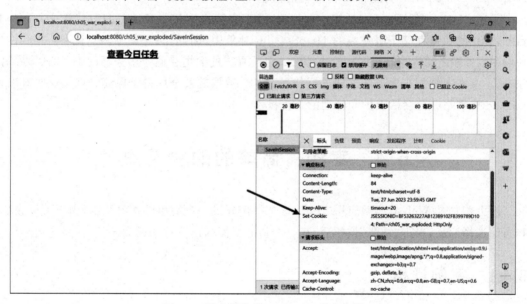

图 5-9　服务端发送 session-id 到客户端

在服务端程序 SaveInSession 中,通过调用 HttpServletRequest 对象的 getSession()方法生成 HttpSession 会话对象,并将 session-id 发送到客户端,如图 5-9 箭头所指向的响应头数据所示。客户端一旦得到了名称为 Set-Cookie 响应头数据,会在后续的 HTTP 请求中携带这个数据到服务端程序。此时,单击"查看今日任务"链接,浏览器将向 ViewSession 服务端程序发送请求,如图 5-10 所示。

如图 5-10 箭头所指向的数据所示,浏览器会将 session-id 发送到服务端程序。服务端程序通过这个 session-id 可以再次获取同一个 HttpSession 对象,然后从中得到存储在

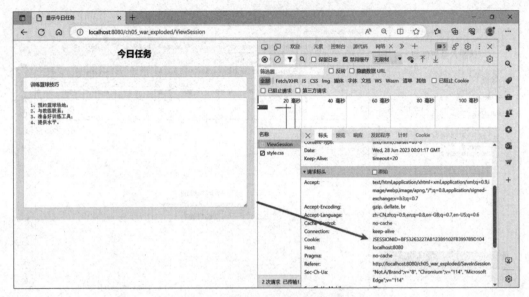

图 5-10　客户端将 session-id 发送到服务端

HttpSession 中的 title 属性和 desc 属性的值。

采用 Session 技术进行会话维护和管理，由于在客户端与服务端之间只传送 session-id 数据，真正的数据存储于服务端的 HttpSession 对象中，因此，具有如下两点优势：其一，客户端和服务端之间传输的数据量较小；其二，由于真正的数据存储于服务端的 HttpSession 对象中，不会在客户端与服务端之间进行传送，所以，数据的安全性得以保证。基于这些有限，Session 技术是进行会话维护和管理的主流技术。

5.4　案例：简单的购物系统

网上购物是 Web 的典型应用之一。这个案例开发一个简单的购物系统，用户可以在购物页面选择商品，单击"结账"按钮后，在一个新的页面列出所选购的商品。

5.4.1　案例目标

设计一个简单的购物系统，在主页面中以优美的方式列出所有可以购买的物品，用户在这个页面中可以选择物品并保存到用户购物车中，当用户单击查看购物车后，在一个新的页面以美观的方式显示用户所选的所有物品信息。

5.4.2　案例分析

这个简单的购物系统应该包括一个商品选购主页面，在这个选购主页面中，用户可以选购想购买的商品。用户选购一件商品后，进入一个新的页面，列出用户所选购的所有商品，此时,用户可以单击"结账"链接进行结算,当然,用户也可以单击"继续购物"链接继续选购商品。因此,这个系统包括如下的程序文件。

(1) shopping.html:主页面,用户在这个主页面中可以选购商品。

(2) shopping.css:页面样式文件。

(3) AddToCart.java:将用户在 shopping.html 页面选购的商品加入购物车。

(4) ListCart.java:当用户单击"结账"链接后进入该页面,列出用户所选购的所有商品。

5.4.3 案例实施

首先编写主页面 shopping.html 代码。编写完成的 shopping.html 的代码如下:

```html
<!DOCTYPE html>
<html lang="en">
<head>
    <meta charset="UTF-8">
    <title>简单的购物系统</title>
    <link href="../css/shopping.css" rel="stylesheet" type="text/css" />
</head>
<body>
    <h2>选择商品,买买买!</h2>
    <div>
        <img src="../images/口哨.jpg"alt=""/>
        <p>口哨,响亮的口哨</p>
        <p>价格:13.8 元</p>
        <p><a href="../AddToCart?m=1">单击购买</a></p>
    </div>
    <div>
        <img src="../images/扩音器.jpg"alt=""/>
        <p>扩音器,个子小声音大</p>
        <p>价格:23.8 元</p>
        <p><a href="../AddToCart?m=2">单击购买</a></p>
    </div>
    <div>
        <img src="../images/拖把.jpg"alt=""/>
        <p>拖把,非常好用的拖把</p>
        <p>价格:45.8 元</p>
        <p><a href="../AddToCart?m=3">单击购买</a></p>
    </div>
    <div>
        <img src="../images/文件柜.jpg"alt=""/>
        <p>文件柜,可以装很多东西</p>
        <p>价格:213.8 元</p>
        <p><a href="../AddToCart?m=4">单击购买</a></p>
    </div>
    <div>
        <img src="../images/沙袋.jpg"alt=""/>
        <p>沙袋,防汛用的沙袋</p>
```

```html
        <p>价格: 53.8元</p>
        <p><a href="../AddToCart?m=5">单击购买</a></p>
    </div>
    <div>
        <img src="../images/灭火器.jpg"alt=""/>
        <p>灭火器,便携式灭火器</p>
        <p>价格: 63.8元</p>
        <p><a href="../AddToCart?m=6">单击购买</a></p>
    </div>
    <div>
        <img src="../images/自救面具.jpg"alt=""/>
        <p>自救面具,紧急情况下使用</p>
        <p>价格: 123.8元</p>
        <p><a href="../AddToCart?m=7">单击购买</a></p>
    </div>
    <div>
        <img src="../images/钥匙.jpg"alt=""/>
        <p>钥匙,非常安全的钥匙</p>
        <p>价格: 13.8元</p>
        <p><a href="../AddToCart?m=8">单击购买</a></p>
    </div>
</body>
</html>
```

在这个页面代码中列出了所有可选购商品。这个页面文件的布局样式 shopping.css 代码如下:

```css
body{
    width: 100%;
    margin: 0  auto;
}
h2{
    text-align: center;
}
div{
    float: left;
    width: 250px;
    height: 380px;
    border: 2px solid rgb(79, 185, 227);
    margin: 10px;
}
img{
    display: block;
    width: 250px;
    height: 250px;
}
```

当在 shopping.html 页面中选择某个商品时,通过如下语句:

```html
<p><a href="../AddToCart?m=8">单击购买</a></p>
```

向服务端程序 AddToCart 发送请求,并将选择商品编号作为参数。AddToCart.java 代码如下：

```java
package com.ttt.servlet;
import jakarta.servlet.ServletConfig;
import jakarta.servlet.ServletException;
import jakarta.servlet.annotation.WebServlet;
import jakarta.servlet.http.HttpServlet;
import jakarta.servlet.http.HttpServletRequest;
import jakarta.servlet.http.HttpServletResponse;
import jakarta.servlet.http.HttpSession;
import java.io.IOException;
import java.util.ArrayList;
import java.util.HashMap;
import java.util.List;
import java.util.Map;
@WebServlet("/AddToCart")
public class AddToCart extends HttpServlet{
    public static Map<String, String> map = new HashMap<>();
    @Override
    public void init(ServletConfig config) throws ServletException{
        super.init(config);
        map.put("1", "口哨");
        map.put("2", "扩音器");
        map.put("3", "拖把");
        map.put("4", "文件柜");
        map.put("5", "沙袋");
        map.put("6", "灭火器");
        map.put("7", "自救面具");
        map.put("8", "钥匙");
    }
    @Override
    protected void doGet(HttpServletRequest req, HttpServletResponse resp)
        throws ServletException, IOException{
        req.setCharacterEncoding("UTF-8");
        String m = req.getParameter("m");
        HttpSession httpSession = req.getSession();
        List<String> things = (ArrayList<String>)httpSession.getAttribute
            ("things");
        if (things == null){
            things = new ArrayList<String>();
            httpSession.setAttribute("things", things);
        }
        things.add(m);
        resp.setContentType("text/html; charset=utf-8");
        resp.getWriter().println ("<p style='text-align: center'>你购买了: " +
                        map.get(m) + "</p>");
        resp.getWriter().println("<p style='text-align: center'>
                    <a href='./html/shopping.html'>继续选购</a></p>");
        resp.getWriter().println("<p style='text-align: center'>
                    <a href='./ListCart'>结账</a></p>");
    }
}
```

在这个代码中,通过以下语句:

```
HttpSession httpSession = req.getSession();
List<String> things = (ArrayList<String>)httpSession.getAttribute("things");
if (things == null){
    things = new ArrayList<String>();
    httpSession.setAttribute("things", things);
}
things.add(m);
```

获得 HttpSession 接口对象,并将新选购的商品加入 HttpSession 接口对象中。当用户单击"结账"链接时,通过访问 ListCart 服务端程序列出用户所选购的所有商品信息。

ListCart.java 的代码如下:

```
package com.ttt.servlet;
import jakarta.servlet.ServletException;
import jakarta.servlet.annotation.WebServlet;
import jakarta.servlet.http.HttpServlet;
import jakarta.servlet.http.HttpServletRequest;
import jakarta.servlet.http.HttpServletResponse;
import jakarta.servlet.http.HttpSession;
import java.io.IOException;
import java.util.ArrayList;
import java.util.List;
@WebServlet("/ListCart")
public class ListCart extends HttpServlet{
    @Override
    protected void doGet(HttpServletRequest req, HttpServletResponse resp)
      throws ServletException, IOException{
        HttpSession httpSession = req.getSession();
        List<String> things = (ArrayList<String>)httpSession.getAttribute
            ("things");
        if ((things == null) || (things.size() ==0)) {
            resp.setContentType("text/html; charset=utf-8");
            resp.getWriter().println("<h2 style='text-align: center'>
                你没有购买任何商品</h2>");
            return;
        }
        resp.setContentType("text/html; charset=utf-8");
        resp.getWriter().println("<h2 style='text-align: center'>你购买如下商品
            </h2>");
        for(String m : things){
            resp.getWriter().println("<p style='text-align: center'>" +
                                AddToCart.map.get(m) + "</p>");
        }
    }
}
```

在这个程序中,通过如下语句:

```
List<String> things = (ArrayList<String>)httpSession.getAttribute("things");
```

从 HttpSession 接口对象中获取用户所选购的商品。然后使用如下语句：

```
for(String m : things) {
    resp.getWriter().println("<p style='text-align: center'>" +
                    AddToCart.map.get(m) + "</p>");
}
```

将用户所选购的商品显示在客户端页面中。运行这个程序，在浏览器地址栏输入 shopping.html，显示如图 5-11 所示界面。

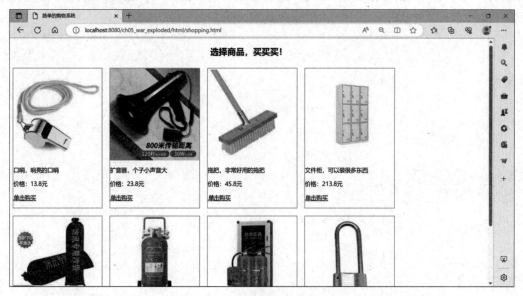

图 5-11　简单的购物系统主页面

在图 5-11 所示的页面中选购商品，然后单击"单击购买"链接，将显示如图 5-12 所示的页面。

图 5-12　购买商品显示页面

用户可以多次选购商品，当在图 5-12 的界面中单击"结账"链接后，将在如图 5-13 所示的界面显示用户选购的所有商品信息。

图 5-13　列出用户所选购的所有商品页面

5.5 练习：记录用户上次登录的时间和地点

在基于 Web 的应用系统中，后台服务系统可以记录用户上一次登录系统的时间和地点，并在用户下次使用系统时告知用户，用户通过这些信息可防止他人冒充和非法登录使用系统。编写一个简单的登录页面，在用户成功登录系统后记录用户的登录的地址和登录时间，并在用户下次登录时显示上一次登录地点和时间。

第 6 章　Servlet 监听器和过滤器

类似于 GUI(graphic user interface,图形用户界面)程序,当用户在图形界面程序中单击某个按钮时,触发一个特定的事件,进而执行预先设置的一个特定程序或函数。在 Java Web 程序中,也可以对系统发生的特定事件进行监听并执行一段预先设置的特定程序或函数,这个对特定事件进行处理的程序称为监听器。在 Java Web 程序中,除了可以监听系统发生的特定事件外,还可以对客户端请求服务端 Servlet 程序的过程进行过滤,并根据应用的需要拦截请求或者在请求上附加特别的数据。

6.1　Servlet 监听器

Servlet 监听器可以对 ServletContext 对象、HttpSession 对象和 HttpServletRequest 对象的创建、销毁及属性变化进行监听。

6.1.1　监听 ServletContext 对象

有两个用于监听 ServletContext 对象的接口:其一,ServletContextListener 接口对 ServletContext 对象的创建和销毁进行监听;其二,ServletContextAttributeListener 接口对 ServletContext 对象中的属性变化进行监听。这两个接口的方法及其描述分别如表 6-1 和表 6-2 所示。

表 6-1　ServletContextListener 接口的方法及描述

序号	方 法 名	描 述
1	default void contextInitialized(ServletContextEvent sce)	当 Servlet 容器初始化 ServletContext 对象时将调用这个方法。也就是说,当 Servlet 容器启动 Java Web 应用程序并为之创建 ServletContext 对象时将调用这个方法
2	default void contextDestroyed(ServletContextEvent sce)	当 Servlet 容器即将销毁 ServletContext 对象时将调用这个方法。也就是说,当 Servlet 容器停止 Java Web 应用程序时将调用这个方法

表 6-2 ServletContextAttributeListener 接口的方法及描述

序号	方法名	描述
1	default void attributeAdded(ServletContextAttributeEvent event)	当向 ServletContext 对象中加入一个新属性时将调用这个方法
2	default void attributeRemoved(ServletContextAttributeEvent event)	当从 ServletContext 对象中移除一个属性时将调用这个方法
3	default void attributeReplaced(ServletContextAttributeEvent event)	当 ServletContext 对象中的一个属性的值被更改时调用这个方法

为了在应用程序监听 ServletContext 对象的创建、销毁及属性变化，需要编写一个实现 ServletContextListener 接口和 ServletContextAttributeListener 接口的类，同时需要使用 @WebListener 注解标注这些类。下面通过一个例子说明如何监听 ServletContext 对象的创建、销毁和属性变化。为此，新建一个名称为 ch06 的 Java Web 工程。新建完成后的工程如图 6-1 所示。

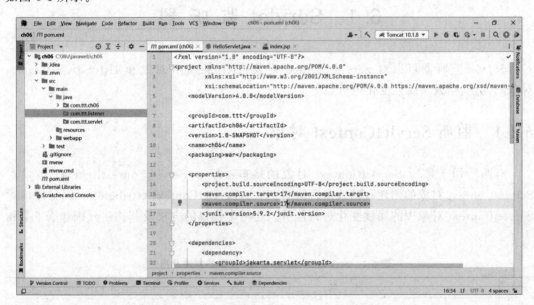

图 6-1 新建的 ch06 Java Web 工程

在 ch06 工程下，新建两个包 com.ttt.servlet 和 com.ttt.listener，分别用于存放 Servlet 程序和监听器程序。首先编写一个简单的 HTML 页面 index.html，这个页面会访问服务端 ServletForListener 程序。index.html 代码如下：

```
<!DOCTYPE html>
<html lang="en">
<head>
    <meta charset="UTF-8">
    <title>Title</title>
</head>
<body>
```

```html
<h2>观察 ServletContext 对象监听器</h2>
<a href="../ServletForListener">单击观察 ServletContext 对象监听器</a>
</body>
</html>
```

在服务端 ServletForListener 程序的 doGet()方法中,在获得了 ServletContext 对象后,向其中添加一些属性,再修改一些属性,最后移除一个属性。ServletForListener.java 代码如下:

```java
package com.ttt.servlet;
import jakarta.servlet.ServletContext;
import jakarta.servlet.ServletException;
import jakarta.servlet.annotation.WebServlet;
import jakarta.servlet.http.HttpServlet;
import jakarta.servlet.http.HttpServletRequest;
import jakarta.servlet.http.HttpServletResponse;
import java.io.IOException;
@WebServlet("/ServletForListener")
public class ServletForListener extends HttpServlet {
    @Override
    protected void doGet(HttpServletRequest req, HttpServletResponse resp)
                            throws ServletException, IOException{
        ServletContext sc = req.getServletContext();
        sc.setAttribute("product", "手电筒");
        sc.setAttribute("price", 23.4f);
        sc.setAttribute("price", 34.5f);
        sc.removeAttribute("product");
        resp.setContentType("text/html; charset=UTF-8");
        resp.getWriter().println("观察添加、修改、移除一些属性后的变化");
    }
}
```

为了监听 ServletContext 对象的被创建(也就是被初始化)、销毁及监听其中的属性值的变化,编写 MyServletContextListenerImpl 类,这个类实现了 ServletContextListener 接口;编写了 MyServletContextAttributeListenerImpl 类,这个类实现了 ServletContextAttributeListener 接口。MyServletContextListenerImpl.java 代码如下:

```java
package com.ttt.listener;
import jakarta.servlet.ServletContextEvent;
import jakarta.servlet.ServletContextListener;
import jakarta.servlet.annotation.WebListener;
@WebListener
public class MyServletContextListenerImpl implements ServletContextListener {
    @Override
    public void contextInitialized(ServletContextEvent sce) {
        ServletContextListener.super.contextInitialized(sce);
        System.out.println("ServletContext Initialized\n");
    }
    @Override
    public void contextDestroyed(ServletContextEvent sce){
        ServletContextListener.super.contextDestroyed(sce);
        System.out.println("ServletContext Destroyed\n");
    }
}
```

这个程序在 ServletContext 对象被创建和被销毁时在控制台显示一段信息，以便可以观察到 ServletContext 被创建的时机和被销毁的时机。注意其中的@WebListener 标注。

所有的 Servlet 监听器都使用这个注解进行标注。MyServletContextAttributeListenerImpl 用以监听 ServletContext 中属性的变化。

MyServletContextAttributeListenerImpl.java 代码如下：

```java
package com.ttt.listener;
import jakarta.servlet.ServletContextAttributeEvent;
import jakarta.servlet.ServletContextAttributeListener;
import jakarta.servlet.annotation.WebListener;
@WebListener
public class MyServletContextAttributeListenerImpl
    implements ServletContextAttributeListener {
    @Override
    public void attributeAdded(ServletContextAttributeEvent event){
        ServletContextAttributeListener.super.attributeAdded(event);
        System.out.println("向 ServletContext 添加了属性");
        System.out.println(event.getName() + ":" + event.getValue() + "\n");
    }
    @Override
    public void attributeRemoved(ServletContextAttributeEvent event) {
        ServletContextAttributeListener.super.attributeRemoved(event);
        System.out.println("从 ServletContext 移除了属性:" + event.getName() + "\n");
    }
    @Override
    public void attributeReplaced(ServletContextAttributeEvent event){
        ServletContextAttributeListener.super.attributeReplaced(event);
        System.out.println("ServletContext 中的属性值发生了变化");
        System.out.println(event.getName() + ",旧值:" + event.getValue() + "新值:" + event.getServletContext().getAttribute(event.getName()) + "\n");
    }
}
```

这个监听器在 ServletContext 对象中的属性发生变化时会被调用。程序在控制台显示一些信息以便观察。现在运行这个程序，观察控制台信息，如图 6-2 所示。

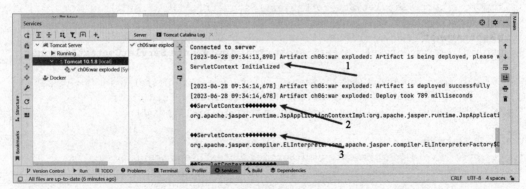

图 6-2 启动程序时 Servlet 容器创建了 ServletContext 对象

第 6 章 Servlet 监听器和过滤器

从图 6-2 可以看出，当启动程序运行后，Servlet 容器创建了 ServletContext 对象，如箭头 1 所指向的信息所示，但是，在图 6-2 的界面中出现了乱码。出现这个乱码的原因是 Tomcat 没有正确配置运行时的字符编码。为了解决这个问题，在图 6-1 的界面中单击下拉框，如图 6-3 所示。

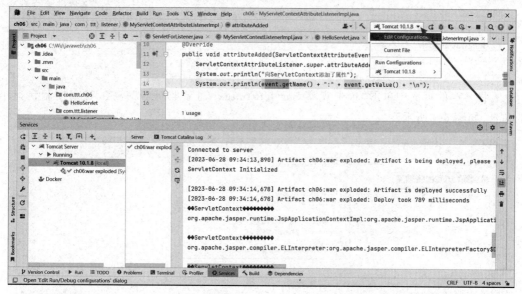

图 6-3　选择配 Tomcat 运行时配置参数

在图 6-3 的界面中，选择"Edit Configurations …"，显示如图 6-4 的界面。

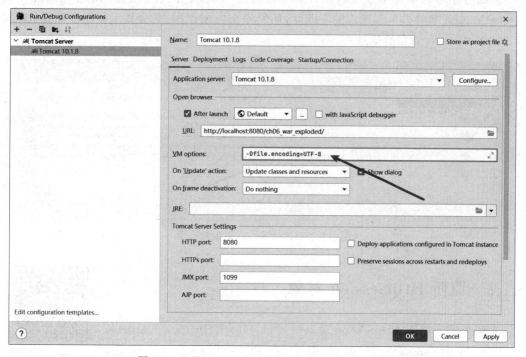

图 6-4　配置 Tomcat 运行时的字符编码为 UTF-8

117

在 VM options 文本框中输入"-Dfile.encoding＝UTF-8",单击 OK 按钮即可,也就是采用 UTF-8 对字符进行编解码。现在再次运行程序,不会再出现乱码。现在在浏览器地址栏输入 index.html,显示如图 6-5 所示的界面。

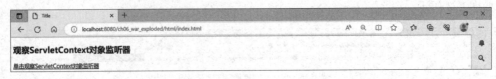

图 6-5　首页 index.html 页面

在图 6-5 的页面中单击"单击观察 ServletContext 对象监听器"链接,观察控制台显示的信息,如图 6-6 所示。

图 6-6　ServletContext 对象的属性变化

从图 6-6 可以清楚看到 ServletContext 对象中的属性变化过程,而这些信息正是 ServletContext 属性监听器输出的。此时,如果停止应用程序,则会显示如图 6-7 所示的销毁 ServletContext 对象的信息。

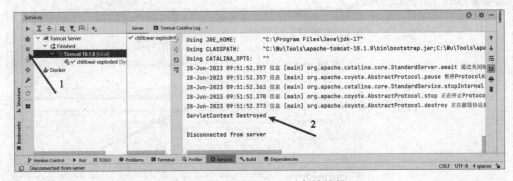

图 6-7　ServletContext 对象被销毁

6.1.2　监听 HttpSession 对象

有两个用于监听 HttpSession 对象的接口:其一,HttpSessionListener 接口用于 HttpSession 对象的创建和销毁进行监听;其二, HttpSessionAttributeListener 接口用于 HttpSession 对象中的属性变化进行

Java Web 程序乱码解决办法总结

监听。这两个接口的方法及其描述分别如表 6-3 和表 6-4 所示。

表 6-3　HttpSessionListener 接口的方法及描述

序号	方法名	描述
1	default void sessionCreated(HttpSessionEvent se)	当创建一个新的 HttpSession 对象时将调用这个方法。也就是说,当应用程序调用 HttpServletRequest 对象的 getSession() 方法创建一个新的 HttpSession 对象时将调用这个方法
2	default void sessionDestroyed(HttpSessionEvent se)	当即将销毁 HttpSession 对象时将调用这个方法。也就是说,当应用程序调用 HttpSession 的 invalidate() 方法时将调用这个方法

表 6-4　HttpSessionAttributeListener 接口的方法及描述

序号	方法名	描述
1	default void attributeAdded (HttpSessionBindingEvent event)	当向 HttpSession 对象中加入一个新属性时将调用这个方法
2	default void attributeRemoved (HttpSessionBindingEvent event)	当从 HttpSession 对象中移除一个属性时将调用这个方法
3	default voidattributeReplaced (HttpSessionBindingEvent event)	当 HttpSession 对象中的一个属性的值被更改时调用这个方法

6.1.3　监听 HttpServletRequest 对象

有两个用于监听 HttpServletRequest 对象的接口：其一,ServletRequestListener 接口用于对 HttpServletRequest 对象的创建和销毁进行监听；其二,ServletRequestAttributeListener 接口用于对 HttpServletRequest 对象中的属性变化进行监听。这两个接口的方法及其描述分别如表 6-5 和表 6-6 所示。

表 6-5　ServletRequestListener 接口的方法及描述

序号	方法名	描述
1	default void requestInitialized(ServletRequestEvent sre)	当 Servlet 容器创建一个新的 HttpServletRequest 对象时将调用这个方法。也就是说,当 Servlet 容器创建一个新的 HttpServletRequest 对象时将调用这个方法
2	default void requestDestroyed(ServletRequestEvent sre)	当 Servlet 容器即将销毁 HttpServletRequest 对象时将调用这个方法

表 6-6　ServletRequestAttributeListener 接口的方法及描述

序号	方法名	描述
1	default void attributeAdded(ServletRequestAttributeEvent srae)	当向 HttpServletRequest 对象中加入一个新属性时将调用这个方法
2	default void attributeRemoved(ServletRequestAttributeEvent srae)	当从 HttpServletRequest 对象中移除一个属性时将调用这个方法
3	default void attributeReplaced(ServletRequestAttributeEvent srae)	当 HttpServletRequest 对象中的一个属性的值被更改时调用这个方法

由于 HttpSession 对象和 HttpServletRequest 对象的创建、销毁，以及属性的变化过程与 ServletContext 对象的创建、销毁，以及属性的变化过程非常类似，仅仅只是时机不同，因此，在此不再举例说明这两个对象的创建、销毁，以及属性的变化过程。

6.2　Filter 过滤器

Filter 过滤器也称为 Servlet Filter 或 Servlet 过滤器，是介于客户端程序与 Servlet 服务端程序之间的一个程序，它会对从客户端发送给服务端 Servlet 程序的请求进行过滤处理。例如，检查请求中是否包含某个属性数据等，极端情况下还可以阻止将请求发送到服务端 Servlet 程序。不仅如此，Servlet 过滤器还可以对服务端 Servlet 程序发送给客户端的应答信息进行处理，例如，在应答中添加特定的信息等。Servlet 过滤器的工作过程如图 6-8 所示。

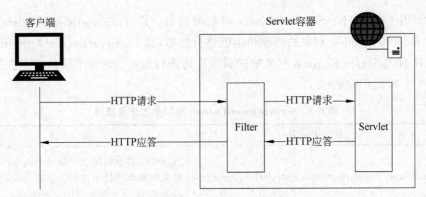

图 6-8　Servlet 过滤器的工作过程

从图 6-8 看出，Servlet 过滤器也运行在 Servlet 容器中，它处于客户端与 Servlet 程序之间，可以对指定的 Servlet 服务端程序的请求进行过滤或者拦截，并且对 Servlet 发送到客户端的 HTTP 响应进行处理。

6.2.1 Filter 接口及其实现类 HttpFilter

为了编写 Servlet 过滤器,需要实现 Filter 接口。Filter 接口定义了三个方法,这三个方法构成了 Filter 的生命周期函数。Filter 接口方法及其描述如表 6-7 所示。

表 6-7 Filter 接口的方法及描述

序号	方 法 名	描 述
1	default void init(FilterConfig filterConfig)	当容器需要使能这个过滤器并使之进入工作状态时,容器将自动调用这个方法
2	void doFilter(ServletRequest request,ServletResponse response,FilterChain chain)	当每次需要过滤器对请求进行过滤处理及对响应进行处理时,容器都会调用过滤器的这个方法实施具体的过滤服务
3	default void destroy()	当容器需要停止这个过滤器的服务时,容器将自动调用这个方法

从 Servlet 标准 4.0 开始,Java Web 提供了 HttpFilter 类。HttpFilter 类实现了 Filter 接口。任何过滤器可以继承这个类并重写 doFilter(ServletRequest request,ServletResponse response,FilterChain chain)方法即可。

为了指定 Filter 对哪个或者哪些 Servlet 服务端程序的访问进行过滤,Java Web 提供了两种配置方式:其一,使用@WebFilter 注解进行配置;其二,在 web.xml 配置文件中配置。由于使用@WebFilter 注解进行配置是比较简洁明了,因此,本书采用@WebFilter 注解对过滤器进行配置。@WebFilter 注解的常用属性及其含义如表 6-8 所示。

表 6-8 @WebFilter 注解的常用属性及其含义

序号	属 性	描 述
1	String filterName	过滤器名称
2	WebInitParam[] initParams	过滤器配置参数,其中每个参数都是 WebInitParam 对象,也就是"名:值"对
3	String[] urlPatterns	过滤器实施过滤的 URL 的模式,例如,模式/a*将对所有以字母 a 开头的 URL 进行过滤处理
4	String[] value	同 urlPatterns 属性,使用时二者只能使用其中一个

例如,使用@WebFilter 配置过滤器的示例如下:

```
@WebFilter(
    filterName = "MyFilter",
    /*通配符(*)表示对所有的web资源进行过滤*/
    urlPatterns = "/*",
    /*这里可以放一些初始化的参数*/
    initParams = {
        @WebInitParam(name = "charset", value = "utf-8")
        @WebInitParam(name = "name", value = "张三")
    }
)
```

6.2.2 Servlet 过滤器应用举例

为了理解 Servlet 过滤器的工作原理及如何编写过滤器程序，下面举例说明如何使用 Servlet 过滤器对 Servlet 服务端程序进行过滤处理。

这个例子中有两个 Servlet：ServletA 和 ServletB，其中，ServletA 从 HTTP 请求中获取名称为 product 的属性，并将这个值显示在浏览器上；ServletB 直接在浏览器上显示 Hello World 文字。两个过滤器 FilterA 和 FilterB 对 HTTP 请求进行过滤处理，其中，FilterA 对访问 ServletA 的请求进行过滤处理，向请求中添加名称为 product 的属性，其值为"空调"，同时，该过滤器还会在 HTTP 响应中添加一个文字"这是 FilterA 添加的文字"。因此，FilterA 在 HTTP 响应中添加的文字将显示在浏览器上；FilterB 则对访问 ServletB 的请求进行过滤，它武断地拦截所有对 ServletB 的请求，也就是说，客户端无法访问到 ServletB 服务端程序。这个例子中的 ServletA、ServletB 和 FilterA、FilterB 的关系如图 6-9 所示。

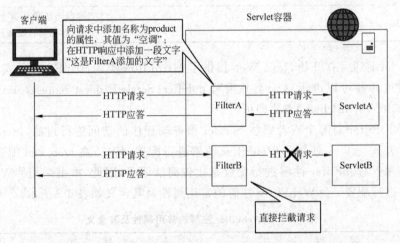

图 6-9 两个过滤器与两个 Servlet 服务程序之间的过滤关系

现在编写 ServletA、ServletB、FilterA、FilterB 的代码。先编写 ServletA 的代码。ServletA 只是从 HTTP 请求中获取属性名为 product 的属性值，并显示在浏览器上。ServletA.java 的代码如下：

```java
package com.ttt.servlet;
import jakarta.servlet.ServletException;
import jakarta.servlet.annotation.WebServlet;
import jakarta.servlet.http.HttpServlet;
import jakarta.servlet.http.HttpServletRequest;
import jakarta.servlet.http.HttpServletResponse;
import java.io.IOException;
@WebServlet("/ServletA")
public class ServletA extends HttpServlet{
    @Override
    protected void doGet(HttpServletRequest req, HttpServletResponse resp)
        throws ServletException, IOException{
        req.setCharacterEncoding("UTF-8");
```

```
        String product = (String)req.getAttribute("product");
        resp.setContentType("text/html; charset=UTF-8");
        resp.getWriter().println("product:" + product + "<br>");
    }
}
```

ServletB.java 的代码如下:

```
package com.ttt.servlet;
import jakarta.servlet.ServletException;
import jakarta.servlet.annotation.WebServlet;
import jakarta.servlet.http.HttpServlet;
import jakarta.servlet.http.HttpServletRequest;
import jakarta.servlet.http.HttpServletResponse;
import java.io.IOException;
@WebServlet("/ServletB")
public class ServletB extends HttpServlet {
    @Override
    protected void doGet(HttpServletRequest req, HttpServletResponse resp)
            throws ServletException, IOException{
        resp.setContentType("text/html; charset=UTF-8");
        resp.getWriter().println("Hello, World!");
    }
}
```

ServletA 和 ServletB 的代码都比较简单,只是向浏览器发送了简单的文字信息。下面编写 FilterA 的代码。按要求,FilterA 要在 HTTP 请求对象中添加名称为 product 的属性,并且还要在 HTTP 的响应中添加"这是 FilterA 添加的文字"这段信息。FilterA.java 代码如下:

```
package com.ttt.filter;
import jakarta.servlet.*;
import jakarta.servlet.annotation.WebFilter;
import java.io.IOException;
@WebFilter(filterName = "MyFilterA", urlPatterns = "/ServletA")
public class FilterA implements Filter{
    @Override
    public void doFilter(ServletRequest req, ServletResponse resp,
        FilterChain chain) throws IOException, ServletException {
        req.setAttribute("product", "空调");
        chain.doFilter(req, resp);
        resp.getWriter().println("这是FilterA添加的文字");
    }
}
```

在 FilterA 中,使用如下语句:

```
@WebFilter(filterName = "MyFilterA", urlPatterns = "/ServletA")
```

配置 FilterA 将对访问 ServletA 的所有请求进行过滤处理。然后使用如下语句:

```
req.setAttribute("product", "空调");
```

向 HTTP 请求中添加了名称为 product 的属性,其值为"空调"。进一步使用如下语句:

```
chain.doFilter(req, resp);
```

将 HTTP 请求通过 FilterChain 传递给下一个过滤器或者某个 Servlet 程序。待 chain.doFilter()函数返回,也就是说,待到 FilterChain 后面的其他程序完成对请求的处理后,FilterA 使用如下语句再向 HTTP 响应中添加了一条文字信息:

```
resp.getWriter().println("这是FilterA添加的文字");
```

这条文字信息将在浏览器中显示出来。下面再来看看 FilterB 过滤器的代码。FilterB.java 代码如下:

```
package com.ttt.filter;
import jakarta.servlet.*;
import jakarta.servlet.annotation.WebFilter;
import java.io.IOException;
@WebFilter(filterName = "MyFilterB", urlPatterns = "/ServletB")
public class FilterB implements Filter{
    @Override
    public void doFilter(ServletRequest req, ServletResponse resp,
      FilterChain chain) throws IOException, ServletException {
        resp.setContentType("text/html; charset=UTF-8");
        resp.getWriter().println("FilterB拦截了所有请求");
    }
}
```

与 FilterA 类似,使用如下语句:

```
@WebFilter(filterName = "MyFilterB", urlPatterns = "/ServletB")
```

指定 FilterB 对访问 ServletB 的所有请求进行过滤。然后使用如下语句:

```
resp.setContentType("text/html; charset=UTF-8");
resp.getWriter().println("FilterB拦截了所有请求");
```

直接向客户端发送了响应信息,而并没有将请求传递给 ServletB 代码,因此,FilterB 拦截了访问 ServletB 的所有请求。

现在运行这个程序,在浏览器地址栏直接输入 ServletA 的 URL 地址来访问 ServletA,运行结果如图 6-10 所示。

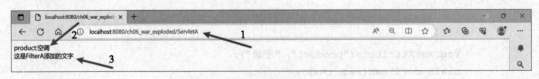

图 6-10 访问 ServletA 的运行结果

从图 6-10 可以看出,ServletA 从请求中获得了 product 属性的值并发送到了客户端,同时 FilterA 在 HTTP 响应中添加了"这是 FilterA 添加的文字"信息。现在在浏览器中直接输入 ServletB 的 URL 地址来访问 ServletB,显示如图 6-11 所示的界面。

从图 6-11 可以看出,FilterB 拦截了访问 ServletB 的所有请求"FilterB 拦截了所有请求"的信息是 FilterB 发送到浏览器中的,而 ServletB 发送的"Hello,World"信息并没有显示在浏览器上,这是因为 ServletB 根本就没有被访问。

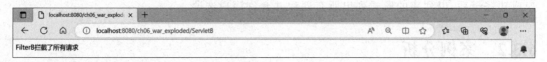

图 6-11　访问 ServletB 的运行结果

6.2.3　FilterChain 接口

服务端程序可以被 0 个、1 个或者多个过滤器处理，这些所有的过滤器形成了一个 FilterChain。客户端的请求和响应正是通过 FilterChain 进行传递的，如图 6-12 所示。

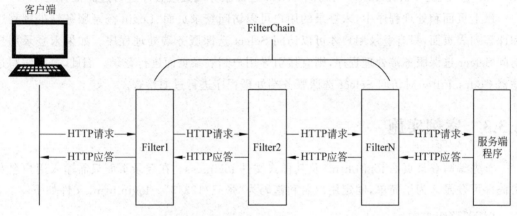

图 6-12　FilterChain

FilterChain 接口中只有一个方法：void doFilter（ServletRequest request，ServletResponse response），处于过滤器链中的过滤器通过调用 FilterChain 接口的 doFilter()方法将请求传递给下一个过滤器，或者当不调用 FilterChain 接口的 doFilter()方法时，则请求不再向后面的过滤器传递，也就是说，对请求做了拦截处理。当然，如果某个过滤器是过滤器链中的最后一个，则调用 FilterChain 接口的 doFilter()方法则是将请求传递给服务端程序。

6.3　案例：使用过滤器检查用户登录状态

在电子商务或者电子政务系统中，在用户可以进行任何操作之前，一般都需要用户登录。可以使用过滤器对用户的请求进行过滤处理，在过滤器中检查用户的登录状态，只有登录的用户才被放行，否则，直接跳转到登录页面。

6.3.1　案例目标

编写一个简单的选课系统，在用户可以选课之前需要用户首先成功登录系统；在用户未成功登录之前，对系统其他页面的访问都将直接将用户引导到登录页面进行登录。在用户成功登录后,以列表的形式列出所有课程并允许用户选课。

6.3.2 案例分析

根据案例目标要求,系统包括如下一些页面和服务端处理程序。
(1) 登录页面:login.html 及其样式文件 login.css。
(2) Login 登录服务端处理程序:对用户的登录请求进行验证。
(3) 课程列表页面:list.html 课程列表页面及其样式文件 list.css。
(4) Select 选课服务端处理程序:对登录用户的选课进行处理。
(5) 过滤器:CourseFilter 对访问 Select 选课服务端处理程序进行过滤检查。

以上页面和处理程序中,未登录的用户可以访问登录页面、Login 登录服务端处理程序和课程列表页面,只有登录用户才可以访问 Select 选课服务端处理程序。如果未登录用户访问 Select 选课服务端处理程序,则直接引导用户到登录页面进行登录。因此,需要编写过滤器 CourseFilter 对访问 Select 选课服务端处理程序进行过滤检查。

6.3.3 案例实施

首先编写登录页面 login.html 及其样式文件 login.css。在登录页面只需输入用户名和密码即可登录。为了简单,固定用户名和密码为"张三/12345"。login.html 文件如下:

```
<!DOCTYPE html>
<html lang="en">
<head>
    <meta charset="utf-8">
    <title>登录</title>
    <link href="../css/login.css" rel="stylesheet" type="text/css" />
</head>
<body>
<div>
    <h3>请填写用户名和密码登录</h3>
    <form method="post"
        enctype="application/x-www-form-urlencoded" action="../Login">
      <label for="username">用户名</label>
      <input type="text" id="username" name="username"
        placeholder="Your name..">
      <label for="password">密码</label>
      <input type="password" id="password"
        name="password" placeholder="Password..">
      <input type="submit" value="Submit">
    </form>
</div>
</body>
</html>
```

其样式文件 login.css 如下:

```css
input[type=text], input[type=password]{
    width: 100%;
    padding: 12px 20px;
    margin: 8px 0;
    display: inline-block;
    border: 1px solid #ccc;
    border-radius: 4px;
    box-sizing: border-box;
}
input[type=submit]{
    width: 100%;
    background-color: #4CAF50;
    color: white;
    padding: 14px 20px;
    margin: 8px 0;
    border: none;
    border-radius: 4px;
    cursor: pointer;
}
input[type=submit]:hover{
    background-color: #45a049;
}
div{
    width: 400px;
    height: 290px;
    margin: 0 auto;
    border-radius: 5px;
    background-color: #f2f2f2;
    padding: 20px;
}
```

当用户在登录页面输入用户名和密码并单击 Submit 按钮后，将登录信息提交给 Login 服务端程序处理。Login.java 的代码如下：

```java
package com.ttt.servlet;
import jakarta.servlet.RequestDispatcher;
import jakarta.servlet.ServletException;
import jakarta.servlet.annotation.WebServlet;
import jakarta.servlet.http.HttpServlet;
import jakarta.servlet.http.HttpServletRequest;
import jakarta.servlet.http.HttpServletResponse;
import java.io.IOException;
@WebServlet("/Login")
public class Login extends HttpServlet{
    @Override
    protected void doPost(HttpServletRequest req, HttpServletResponse resp)
        throws ServletException, IOException{
        req.setCharacterEncoding("UTF-8");
        String un = req.getParameter("username");
        String pass = req.getParameter("password");
        if ((un == null) || (un.isBlank()) || (pass == null) || (pass.isBlank()))
```

```
        {
            resp.setContentType("text/html; charset=UTF-8");
            resp.getWriter().println("<h2 style='text-align: center'>" +
                    "用户名/密码不能为空</h2>");
            return;
        }
        if((!un.equalsIgnoreCase("张三")) ||
            (!pass.equalsIgnoreCase("12345"))) {
            resp.setContentType("text/html; charset=UTF-8");
            resp.getWriter().println("<h2 style='text-align: center'>" +
                    "用户名/密码不正确</h2>");
            return;
        }
        req.getSession();
        resp.sendRedirect("./html/list.html");
    }
}
```

在 Login 服务端代码中,首先判断用户名/密码是否正确,如果正确,则创建 Session 会话,并跳转到课程列表页面 list.html。课程列表页面 list.html 的代码如下:

```
<!DOCTYPE html>
<html lang="en">
<head>
    <meta charset="UTF-8">
    <title>选课系统课程页面</title>
    <link href="../css/list.css" rel="stylesheet" type="text/css" />
</head>
<body>
<h2>课程列表</h2>
<div class="course">
    <img src="../images/prog.png" alt=""/>
    <p>面向对象程序设计</p>
    <p>课时数:72 课时</p>
    <p><a href="../sec/Select?m=面向对象程序设计">单击选课</a></p>
</div>
<div class="course">
    <img src="../images/database.png" alt=""/>
    <p>数据库原理</p>
    <p>课时数:56 课时</p>
    <p><a href="../sec/Select?m=数据库原理">单击选课</a></p>
</div>
<div class="course">
    <img src="../images/stru.png" alt=""/>
    <p>数据结构</p>
    <p>课时数:60 课时</p>
    <p><a href="../sec/Select?m=数据结构">单击选课</a></p>
</div>
<div class="course">
    <img src="../images/h5.png" alt=""/>
```

```html
    <p>HTML5</p>
    <p>课时数:40课时</p>
    <p><a href="../sec/Select?m=HTML5">单击选课</a></p>
</div>
<div class="course">
    <img src="../images/swe.png" alt=""/>
    <p>软件工程</p>
    <p>课时数:60课时</p>
    <p><a href="../sec/Select?m=软件工程">单击选课</a></p>
</div>
</body>
</html>
```

其对应的样式文件 list.css 内容如下:

```css
body{
    width: 900px;
    margin: 0  auto;
}
h2{
    text-align: center;
}
div{
    float: left;
    width: 150px;
    height: 250px;
    border: 2px solid rgb(79, 185, 227);
    margin: 10px;
}
img{
    display: block;
    width: 128px;
    height: 128px;
}
```

在课程列表页面上,用户可以选择课程。当用户选择一门课程时,程序将所选择的课程的课程名称发送到服务端程序 Select 进行处理。Select.java 代码如下:

```java
package com.ttt.servlet;
import jakarta.servlet.ServletException;
import jakarta.servlet.annotation.WebServlet;
import jakarta.servlet.http.HttpServlet;
import jakarta.servlet.http.HttpServletRequest;
import jakarta.servlet.http.HttpServletResponse;
import jakarta.servlet.http.HttpSession;
import java.io.IOException;
@WebServlet("/sec/Select")
public class Select extends HttpServlet{
    @Override
    protected void doGet(HttpServletRequest req, HttpServletResponse resp)
      throws ServletException, IOException{
```

```
        req.setCharacterEncoding("UTF-8");
        String course = req.getParameter("m");
        HttpSession hs = req.getSession(false);
        hs.setAttribute(course, true);
        resp.setContentType("text/html; charset=UTF-8");
        resp.getWriter().println("<h2 style='text-align:center'>选择了: "+
            course+".</h2>");
    }
}
```

注意：采用语句@WebServlet("/sec/Select")将 Select 的 URL 映射到"/sec/Select"下。之所以这么映射，是便于在过滤器 CourseFilter 代码中定义过滤模式参数。

CourseFilter.java 代码如下：

```
package com.ttt.filter;
import jakarta.servlet.*;
import jakarta.servlet.annotation.WebFilter;
import jakarta.servlet.http.HttpServletRequest;
import jakarta.servlet.http.HttpServletResponse;
import java.io.IOException;
@WebFilter(filterName = "CourseFilter", urlPatterns = "/sec/*")
public class CourseFilter implements Filter{
    @Override
    public void doFilter(ServletRequest req, ServletResponse resp, FilterChain chain)
      throws IOException, ServletException{
        HttpServletRequest httpServletRequest = (HttpServletRequest)req;
        if (httpServletRequest.getSession(false) == null){
            String p = httpServletRequest.getContextPath();
            ((HttpServletResponse) resp).sendRedirect(p + "/html/login.html");
            return;
        }
        chain.doFilter(req, resp);
    }
}
```

在过滤器代码中，通过语句@WebFilter(filterName = "CourseFilter"，urlPatterns = "/sec/*")定义对于"/sec"URL 模式下的所有资源访问都要经过 CourseFilter 过滤处理。该过滤器通过如下语句：

```
if (httpServletRequest.getSession(false) == null) {
    String p = httpServletRequest.getContextPath();
    ((HttpServletResponse) resp).sendRedirect(p + "/html/login.html");
    return;
}
```

检查 HttpServletRequest 对象中是否存在 Session，如果存在，则说明用户已经登录，可以进行选课操作，因此，该过滤器对用户直接放行；否则，对于未登录用户，则直接跳转到登录页面。运行这个程序，在浏览器地址栏输入 login.html，显示如图 6-13 所示的界面。

在图 6-13 的页面中输入用户名"张三"和密码"12345"后登录，进入课程列表页面，如图 6-14 所示。

图 6-13　选课登录页面

图 6-14　课程列表页面

选择一门课程,系统会提示用户选课成功,如图 6-15 所示。

图 6-15　选课成功提示页面

现在,关闭浏览器后再打开浏览器(备注:这个操作是为了确保 Session 被服务器关闭),然后在浏览器地址栏直接输入课程列表 list.html 的地址,显示与图 6-14 一样。

可是,如果此时单击"选课选课"链接选择一门课程会发现:未登录用户不能选课,而是直接跳转到登录页面,如图 6-16 所示。这就是过滤器处理的结果。

图 6-16　未登录用户选择课程将直接导向登录页面

6.4 练习：选班长

某班要选班长，有三个班长候选人：张三、李四和王二，在用户可以选择班长之前需要登录到选班长系统。在用户未成功登录之前，对系统任何其他页面的访问都将直接将用户引导到登录页面进行登录。在用户成功登录后，用户通过单选方式选择一名同学为班长。在选班长活动结束后，用户可以查看三个班长候选人的得票统计结果。

提示：使用 ServletContext 保存三个候选人的得票数目。

第 7 章 访问数据库

当程序数据达到一定的规模时,使用数据库对程序数据进行管理是必然趋势。Java Web 程序访问数据库与 Java SE 访问数据库的技术方式是一致的,本质上都是使用 JDBC 访问数据库。但是,由于在 Java Web 程序中,同时访问系统的用户数量可能会非常巨大,因此,对访问数据库技术做了进一步优化,即使用数据库连接池来专门管理程序与数据库管理系统的连接,从而提高程序运行性能。本章对在 Java Web 程序使用 JDBC 访问数据库进行简单介绍后,深入介绍如何使用数据库连接池来提高数据库访问性能。本章采用 MySQL 8.x 作为目标数据库系统。

7.1 使用 JDBC 访问数据库

JDBC(Java Data Base Connectivity,Java 数据库连接),是 Java 提供的访问数据库的统一框架和标准接口。通过 JDBC,Java 程序可以访问和操作任何常见的数据库系统。

7.1.1 使用 JDBC 访问数据库的一般过程

JDBC 接口有效屏蔽了数据库管理系统实现上的差异,从而隔离了应用程序与所使用的数据库管理系统之间的差异,提升了程序的适应性。例如,在开发 Java 应用程序的某个阶段可以使用 MySQL 数据库,在不需要或者只需对程序做较小修改的情况即可访问 SQL Server 数据库,为程序开发带来了便利。

由于 JDBC 是一组标准的 Java 接口,因此,不论使用何种类型的数据库,访问和操作数据库中数据的过程都是一致的,包括如下几个步骤:第 1 步,加载数据库驱动程序;第 2 步,注册数据库驱动程序;第 3 步,获取与数据库的连接;第 4 步,访问和操作数据库中的数据;第 5 步,释放资源并关闭与数据库的连接。

7.1.2 使用 JDBC 访问数据库示例

下面举例说明如何使用 JDBC 访问数据库数据。

首先在 MySQL 8.x 数据库管理系统中创建一个名为 dbschool 的数据库,然后,在其中创建一个名为 student 的表。创建 dbschool 数据库和创建 student 表的语句如下:

```
//创建数据库 dbschool
CREATE SCHEMA'dbschool' DEFAULT CHARACTER SET utf8mb4;
```

```sql
//创建表 student
CREATE TABLE 'dbschool'.'student'(
  'id' INT NOT NULL AUTO_INCREMENT,
  'name' VARCHAR(45) NOT NULL,
  'address' VARCHAR(200) NOT NULL,
  'birth' DATE NOT NULL,
  'entryscore' FLOAT NOT NULL,
  PRIMARY KEY ('id'))
ENGINE = InnoDB
DEFAULT CHARACTER SET = utf8mb4;
```

然后使用如下语句向 student 表中插入一些记录：

```sql
INSERT INTO 'dbschool'.'student' ('name', 'address', 'birth', 'entryscore')
VALUES ('张三', '广东省广州市', '2000-10-10', '650.5');
INSERT INTO 'dbschool'.'student' ('name', 'address', 'birth', 'entryscore')
VALUES ('李四', '北京市海淀区', '2001-01-09', '710');
INSERT INTO 'dbschool'.'student' ('name', 'address', 'birth', 'entryscore')
VALUES ('王二', '广东省珠海市拱北区', '2002-10-20', '700.5');
INSERT INTO 'dbschool'.'student' ('name', 'address', 'birth', 'entryscore')
VALUES ('马六', '河南省郑州市', '2001-12-21', '671');
```

这个例子读取 student 表中的所有记录，并显示在页面上。为此，新建名为 ch07 的 Java Web 工程。由于在新建的工程中，index.jsp 和 HelloServlet 程序是无关紧要的，因此，完全可以将这两个文件删除，甚至可以将 com.ttt.ch07 的包也删除。完成后的 ch07 工程如图 7-1 所示。

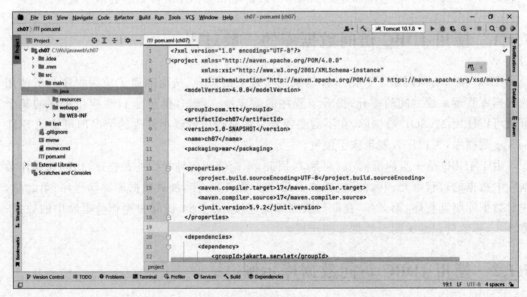

图 7-1　新建名称为 ch07 的 Java Web 工程

在完成工程创建后，需要在 pom.xml 文件中加入 MySQL 的 JDBC 驱动程序。为此，在 pom.xml 文件的<dependencies>标签下加入如下代码：

```xml
<dependency>
    <groupId>com.mysql</groupId>
```

```xml
        <artifactId>mysql-connector-j</artifactId>
        <version>8.0.32</version>
</dependency>
```

现在，编写查询 student 表中所有记录并在页面列表显示的程序。这是一个服务端 Servlet 程序，命名该程序为 StudentMan。StudentMan.java 代码如下：

```java
package com.ttt.servlet;
import jakarta.servlet.ServletException;
import jakarta.servlet.annotation.WebServlet;
import jakarta.servlet.http.HttpServlet;
import jakarta.servlet.http.HttpServletRequest;
import jakarta.servlet.http.HttpServletResponse;
import java.io.IOException;
import java.io.PrintWriter;
import java.sql.*;
@WebServlet("/StudentMan")
public class StudentMan extends HttpServlet{
    Connection conn = null;
    @Override
    public void init() throws ServletException{
        super.init();
        try{
            Class.forName("com.mysql.cj.jdbc.Driver");
            conn = DriverManager.getConnection(
                "jdbc:mysql://localhost:3306/dbschool?" +
                "useSSL=true&useUnicode=true&characterEncoding=utf-8&" +
                "serverTimezone=Asia/Shanghai", "root", "12345");
        } catch (ClassNotFoundException | SQLException e){
            e.printStackTrace();
            log(e.getMessage());
        }
    }
    @Override
    protected void doGet(HttpServletRequest req, HttpServletResponse resp)
      throws ServletException, IOException{
        resp.setContentType("text/html; charset=UTF-8");
        PrintWriter w = resp.getWriter();
        try{
            Statement stat = conn.createStatement();
            ResultSet rs = stat.executeQuery("SELECT * FROM student");
            while(rs.next()){
                w.println("<p style='text-align:center; margin: 10px'>");
                String sb = rs.getInt("id") + "   " +
                    rs.getString("name") + "   " +
                    rs.getString("address") + "   " +
                    rs.getDate("birth") + "   " +
                    rs.getFloat("entryscore") + "   " +
                    "<a href='./DeleteStudent?id=" + rs.getInt("id") +
                    "'>" + "删除" + "</a>";
                w.println(sb);
```

```
            w.println("</p>");
        }
        rs.close();
        stat.close();
    } catch (SQLException e) {
        w.println(e.getMessage());
        log(e.getMessage());
    }
}
```

在 StudentMan 服务端程序的 init()方法中使用如下语句：

```
Class.forName("com.mysql.cj.jdbc.Driver");
conn = DriverManager.getConnection(
        "jdbc:mysql://localhost:3306/dbschool?" +
        "useSSL=true&useUnicode=true&characterEncoding=utf-8&" +
        "serverTimezone=Asia/Shanghai", "root", "12345");
```

加载 JDBC 驱动程序，然后调用 DriverManager 的 getConnection()方法获得一个与数据库的连接，并且将数据库连接保存到 conn 对象变量中。当用户访问 StudentMan 程序时，使用如下语句：

```
Statement stat = conn.createStatement();
ResultSet rs = stat.executeQuery("select * from student");
```

创建 Statement 对象，执行查询 student 表所有数据的 SQL 语句并将查询结果存储在 ResultSet 对象中，进而通过如下语句：

```
String sb = rs.getInt("id") + " " +
        rs.getString("name") + " " +
        rs.getString("address") + " " +
        rs.getDate("birth") + " " +
        rs.getFloat("entryscore") + " " +
        "<a href='./DeleteStudent?id=" + rs.getInt("id") +
        "'>" + "删除" + "</a>";
```

获取每条数据记录，并组装成一个字符串，再通过如下语句：

```
w.println(sb);
w.println("</p>");
```

将记录显示在客户端浏览器上。对于每条记录，通过单击"删除"链接，将访问 DeleteStudent 服务端程序并执行删除操作。

DeleteStudent.java 的代码如下：

```
package com.ttt.servlet;
import jakarta.servlet.ServletException;
import jakarta.servlet.annotation.WebServlet;
import jakarta.servlet.http.HttpServlet;
import jakarta.servlet.http.HttpServletRequest;
import jakarta.servlet.http.HttpServletResponse;
import java.io.IOException;
```

```java
import java.io.PrintWriter;
import java.sql.*;
@WebServlet("/DeleteStudent")
public class DeleteStudent extends HttpServlet{
    Connection conn = null;
    @Override
    public void init() throws ServletException{
        super.init();
        try{
            Class.forName("com.mysql.cj.jdbc.Driver");
            conn = DriverManager.getConnection(
                "jdbc:mysql://localhost:3306/dbschool?" +
                "useSSL=true&useUnicode=true&characterEncoding=utf-8&" +
                "serverTimezone=Asia/Shanghai", "root", "12345");
        } catch (ClassNotFoundException | SQLException e) {
            e.printStackTrace();
            log(e.getMessage());
        }
    }
    @Override
    protected void doGet(HttpServletRequest req, HttpServletResponse resp)
      throws ServletException, IOException{
        req.setCharacterEncoding("UTF-8");
        String id = req.getParameter("id");
        resp.setContentType("text/html; charset=UTF-8");
        PrintWriter w = resp.getWriter();
        if ((id == null) || (id.isBlank())) {
            w.println("<h2 style='text-align:center'>");
            w.println("参数错误</h2>");
            return;
        }
        try{
            PreparedStatement ps = conn.prepareStatement(
                "DELETE FROM student WHERE id = ?");
            ps.setInt(1, Integer.parseInt(id));
            int count = ps.executeUpdate();
            if (count > 0){
                w.println("<h2 style='text-align:center'>");
                w.println("删除了 id=" + id + "的学生信息</h2>");
            }
            ps.close();
        } catch (SQLException e) {
            w.println(e.getMessage());
            log(e.getMessage());
        }
    }
}
```

与 StudentMan 类似，在 DeleteStudent 的 init()方法中，加载 JDBC 驱动程序并且获得一个与数据库连接。在其 doGet()方法中，首先获得要删除记录的 id 参数，然后，创建

PreparedStatement 对象,通过该对象执行删除指定 id 的 student 记录数据。最后,将执行结果显示在客户端。

现在运行程序,在浏览器地址栏输入 StudentMan 的 URL 地址,将显示如图 7-2 所示的界面。

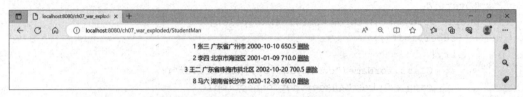

图 7-2　列表 student 表的数据记录

在图 7-2 所示的界面中,单击某个学生信息后面的"删除"链接,则可以从数据库中删除选定的学生信息,如图 7-3 所示。

图 7-3　删除指定 id 的学生信息

7.2　数据库连接池

在 7.1 节的例子中,每个服务端的 Servlet 程序在其 init()方法中都加载了数据库驱动程序并且建立一个数据库的连接,这种方法在并发用户量不大的情况下是可以正常工作的。但是,这种方式存在至少三个潜在问题:其一,数据库连接是重要的资源,每个数据库管理系统能够支持的连接数是有限的;其二,建立数据库连接是一个费时费资源的过程,应该尽量复用数据库连接资源;其三,在并发用户量较大时,由于每个 Servlet 程序都是以单例多线程方式运行的,多个线程会共享一个数据库连接,存在数据库事务不完整的风险。因此,需要使用更好的方法来解决这些问题,这个方法就是使用数据库连接池。

7.2.1　什么是数据库连接池

既然数据库连接是一种资源,那么应该尽可能地充分利用。一个简单而有效的方式就是:可以预先创建多个数据库连接以备用,任何需要使用数据库连接的程序可以从这一组连接中申请一个连接,得到数据库连接对象后即可使用,一旦完成了数据库操作而不再需要数据库连接时,再将这个数据库连接退回,其他程序又可以申请继续使用。这是一个不错的想法,其实这就是数据库连接池的思想。通过使用数据库连接池技术,一方面提高了数据库连接的使用效率;另一方面,由于程序可以对数据库连接"拿来即用",省去了建立数据库连接的时间,程序的运行效率也提高了。

7.2.2 DataSource 接口

Java 为数据库连接池专门定义了 DataSource 接口。DataSource 接口中有一个非常重要的方法为 Connection getConnection()，通过这个方法可以从 DataSource 接口中获得一个数据库连接对象。程序一旦通过这个方法得到一个数据库连接对象，可以像使用普通数据库连接对象一样使用它来操作数据库数据。

因为 DataSource 只是一个接口，因此，需要一个类来实现这个接口，进而可以通过这个接口来管理数据库连接对象。目前常用的、实现了 DataSource 接口的实现类包括 DBCP、C3P0、BoneCP、Proxool、Druid 等。由于这些数据库连接池的使用方式比较相似，因此，下面对 DBCP 和 Druid 的使用进行介绍。

7.2.3 使用 DBCP 建立数据库连接池

DBCP(Data Base Connection Pool，数据库连接池)是 Apache 开源组织旗下的一个数据库连接池开源软件。为了在程序中使用 DBCP，需要在 pom.xml 文件中导入如下所示的依赖：

```xml
<dependency>
    <groupId>org.apache.commons</groupId>
    <artifactId>commons-dbcp2</artifactId>
    <version>2.9.0</version>
</dependency>
```

DBCP 有两个核心类为 BasicDataSource 类和 BasicDataSourceFactory 类，其中，BasicDataSource 类实现了 DataSource 接口，而 BasicDataSourceFactory 类则在 BasicDataSource 类的基础上进行了进一步的封装，可以从配置文件中读取创建数据库连接池的相关参数。现阶段，更为常用的方式是使用配置文件进行数据库连接相关参数配置。下面举例说明如何使用配置文件和 BasicDataSourceFactory 类创建数据库连接池。还是以 7.2 节操作 student 表中的数据为例说明如何使用 DBCP 建立数据库连接池。

首先在 ch07 工程的 source/main/resources 目录下新建名为 dbcp.properties 的配置文件，其内容如下：

```
driverClassName=com.mysql.cj.jdbc.Driver
url=jdbc:mysql://localhost:3306/dbschool?serverTimezone=Asia/Shanghai
username=root
password=12345
initialSize=5
maxIdle=10
```

通过这个配置文件，告知了 DBCP 有关数据库驱动程序、连接 URL 串、登录数据库的用户名密码、初始连接数及最大空闲连接数等信息。DBCP 将使用这些配置信息创建与指定数据库的连接池。

在编程实践中，一般都会创建一个专门的工具类来管理数据库连接池。为此，创建一个专门的 package 来存放这些工具类,命名这个包为 com.ttt.utils。在这个包下创建名称为 DBCPUtil 的工具类。DBCPUtil.java 的代码如下:

```java
package com.ttt.utils;
import org.apache.commons.dbcp2.BasicDataSourceFactory;
import javax.sql.DataSource;
import java.io.InputStream;
import java.sql.Connection;
import java.sql.SQLException;
import java.util.Properties;
public class DBCPUtil{
    private static DataSource ds = null;
    public static Connection getConnection(){
        if (ds == null){
            InputStream in = DBCPUtil.class.getClassLoader().
                getResourceAsStream("dbcp.properties");
            Properties prop = new Properties();
            try{
                prop.load(in);
                ds = BasicDataSourceFactory.createDataSource(prop);
            }
            catch (Exception e){
                throw new RuntimeException(e);
            }
        }
        try{
            return ds.getConnection();
        }
        catch (SQLException e){
            throw new RuntimeException(e);
        }
    }
}
```

在 DBCPUtil 程序中,使用如下语句:

```java
InputStream in = DBCPUtil.class.getClassLoader().getResourceAsStream("dbcp.properties");
Properties prop = new Properties();
```

读取配置文件信息。然后使用如下语句:

```java
prop.load(in);
ds = BasicDataSourceFactory.createDataSource(prop);
```

创建了 DataSource 接口,进而,可从 DataSource 接口中通过使用 getConnection() 方法获取一个数据库连接。

现在修改 StudentMan 服务端程序并重新命名为 StudentManDBCP。StudentManDBCP.java 程序的代码如下:

```java
package com.ttt.servlet;
import com.ttt.utils.DBCPUtil;
import jakarta.servlet.ServletException;
```

```java
import jakarta.servlet.annotation.WebServlet;
import jakarta.servlet.http.HttpServlet;
import jakarta.servlet.http.HttpServletRequest;
import jakarta.servlet.http.HttpServletResponse;
import java.io.IOException;
import java.io.PrintWriter;
import java.sql.*;
@WebServlet("/StudentManDBCP")
public class StudentManDBCP extends HttpServlet{
    @Override
    protected void doGet(HttpServletRequest req, HttpServletResponse resp)
        throws ServletException, IOException{
        resp.setContentType("text/html; charset=UTF-8");
        PrintWriter w = resp.getWriter();
        try{
            Connection conn = DBCPUtil.getConnection();
            Statement stat = conn.createStatement();
            ResultSet rs = stat.executeQuery("select * from student");
            while(rs.next()){
                w.println("<p style='text-align:center; margin: 10px'>");
                String sb = rs.getInt("id") + " " +
                            rs.getString("name") + " " +
                            rs.getString("address") + " " +
                            rs.getDate("birth") +" " +
                            rs.getFloat("entryscore") + " " +
                            "< a href = './DeleteStudentDBCP? id=" + rs.getInt("id") +
                            "'>" + "删除" + "</a>";
                w.println(sb);
                w.println("</p>");
            }
            rs.close();
            stat.close();
            conn.close();
        } catch (SQLException e){
            w.println(e.getMessage());
            log(e.getMessage());
        }
    }
}
```

在 StudentManDBCP 程序中，使用语句"Connection conn = DBCPUtil.getConnection();"从数据库连接池获取一个连接，然后在该连接上执行数据库相关操作。最后使用语句"conn.close();"将数据库连接归还给数据库连接池。

与此类似，对 DeleteStudent 代码进行修改，并重新命名为 DeleteStudentDBCP。DeleteStudentDBCP.java 代码如下：

```java
package com.ttt.servlet;
import com.ttt.utils.DBCPUtil;
import jakarta.servlet.ServletException;
```

```java
import jakarta.servlet.annotation.WebServlet;
import jakarta.servlet.http.HttpServlet;
import jakarta.servlet.http.HttpServletRequest;
import jakarta.servlet.http.HttpServletResponse;
import java.io.IOException;
import java.io.PrintWriter;
import java.sql.Connection;
import java.sql.DriverManager;
import java.sql.PreparedStatement;
import java.sql.SQLException;
@WebServlet("/DeleteStudentDBCP")
public class DeleteStudentDBCP extends HttpServlet{
    @Override
    protected void doGet(HttpServletRequest req, HttpServletResponse resp)
      throws ServletException, IOException{
        req.setCharacterEncoding("UTF-8");
        String id = req.getParameter("id");
        resp.setContentType("text/html; charset=UTF-8");
        PrintWriter w = resp.getWriter();
        if ((id == null) || (id.isBlank())) {
            w.println("<h2 style='text-align:center'>");
            w.println("参数错误</h2>");
            return;
        }
        try{
            Connection conn = DBCPUtil.getConnection();
            PreparedStatement ps = conn.prepareStatement(
                "DELETE FROM from student WHERE id = ?");
            ps.setInt(1, Integer.parseInt(id));
            int count = ps.executeUpdate();
            if (count > 0){
                w.println("<h2  style='text-align:center'>");
                w.println("删除了 id=" + id + "的学生信息</h2>");
            }
            ps.close();
            conn.close();
        } catch (SQLException e){
            w.println(e.getMessage());
            log(e.getMessage());
        }
    }
}
```

现在运行程序，在浏览器地址栏输入 StudentManDBCP 的 URL 地址，将得到与图 7-2 一致的结果。

7.2.4 使用 Druid 建立数据库连接池

Druid 是目前比较流行的高性能数据库连接池，是阿里的开源产品，提供了快速的数据聚合能力以及亚秒级的查询能力。在使用 Druid 之前，需要在 pom.xml 文件中导入相关依赖：

```xml
<dependency>
    <groupId>com.alibaba</groupId>
    <artifactId>druid</artifactId>
    <version>1.2.18</version>
</dependency>
```

与 DBCP 类似，Druid 有两个核心类为 DruidDataSource 类和 DruidDataSourceFactory 类。由于使用配置文件创建数据库连接池是主要的使用方式，因此，在 ch07 工程的 src/main/resources 目录下新建 druid.properties 文件，文件内容如下：

```
driverClassName=com.mysql.cj.jdbc.Driver
url=jdbc:mysql://localhost:3306/dbschool?serverTimezone=Asia/Shanghai
username=root
password=12345
initialSize=10
maxActive=50
minIdle=5
maxWait=6000
```

与 DBCP 类似，为 Druid 数据库连接池创建一个专门的工具类 DruidUtil。DBCPUtil.java 的代码如下：

```java
package com.ttt.utils;
import com.alibaba.druid.pool.DruidDataSourceFactory;
import javax.sql.DataSource;
import java.io.InputStream;
import java.sql.Connection;
import java.sql.SQLException;
import java.util.Properties;
public class DruidUtil{
    private static DataSource ds = null;
    public static Connection getConnection(){
        if (ds == null){
            InputStream in = DruidUtil.class.getClassLoader().
                getResourceAsStream("druid.properties");
            Properties prop = new Properties();
            try{
                prop.load(in);
                ds = DruidDataSourceFactory.createDataSource(prop);
            } catch (Exception e){
                throw new RuntimeException(e);
            }
        }
        try{
            return ds.getConnection();
        } catch (SQLException e){
            throw new RuntimeException(e);
        }
    }
}
```

除了使用不同的数据库连接池工厂类不同外，DruidUtil 与 DBCPUtil 程序是完全一致的。在 Druid 中，使用如下语句所示的 Druid 数据库连接工厂类建立数据库连接池：

ds = DruidDataSourceFactory.createDataSource(prop);

现在修改 StudentMan 服务端程序并重新命名为 StudentManDruid。StudentManDruid.java 程序的代码如下：

```java
package com.ttt.servlet;
import com.ttt.utils.DBCPUtil;
import com.ttt.utils.DruidUtil;
import jakarta.servlet.ServletException;
import jakarta.servlet.annotation.WebServlet;
import jakarta.servlet.http.HttpServlet;
import jakarta.servlet.http.HttpServletRequest;
import jakarta.servlet.http.HttpServletResponse;
import java.io.IOException;
import java.io.PrintWriter;
import java.sql.Connection;
import java.sql.ResultSet;
import java.sql.SQLException;
import java.sql.Statement;
@WebServlet("/StudentManDruid")
public class StudentManDruid extends HttpServlet{
    @Override
    protected void doGet(HttpServletRequest req, HttpServletResponse resp)
      throws ServletException, IOException{
        resp.setContentType("text/html; charset=UTF-8");
        PrintWriter w = resp.getWriter();
        try{
            Connection conn = DruidUtil.getConnection();
            Statement stat = conn.createStatement();
            ResultSet rs = stat.executeQuery("SELECT * FROM student");
            while(rs.next()){
                w.println("<p style='text-align:center; margin: 10px'>");
                String sb = rs.getInt("id") + "  " +
                        rs.getString("name") + "  " +
                        rs.getString("address") + "  " +
                        rs.getDate("birth") + "  " +
                        rs.getFloat("entryscore") + "  " +
                        "<a href='./DeleteStudentDruid?id=" + rs.getInt
                        ("id") + "'>" + "删除" + "</a>";
                w.println(sb);
                w.println("</p>");
            }
            rs.close();
            stat.close();
            conn.close();
        } catch (SQLException e){
            w.println(e.getMessage());
```

```
            log(e.getMessage());
        }
    }
}
```

除了使用不同的数据库连接池工具类外，StudentManDruid 程序与 StudentManDBCP 程序是一致的。与此类似，对 DeleteStudent 代码进行修改，并重新命名为 DeleteStudentDruid。DeleteStudentDruid.java 代码如下：

```
package com.ttt.servlet;
import com.ttt.utils.DBCPUtil;
import com.ttt.utils.DruidUtil;
import jakarta.servlet.ServletException;
import jakarta.servlet.annotation.WebServlet;
import jakarta.servlet.http.HttpServlet;
import jakarta.servlet.http.HttpServletRequest;
import jakarta.servlet.http.HttpServletResponse;
import java.io.IOException;
import java.io.PrintWriter;
import java.sql.Connection;
import java.sql.PreparedStatement;
import java.sql.SQLException;
@WebServlet("/DeleteStudentDruid")
public class DeleteStudentDruid extends HttpServlet{
    @Override
    protected void doGet(HttpServletRequest req, HttpServletResponse resp)
      throws ServletException, IOException{
        req.setCharacterEncoding("UTF-8");
        String id = req.getParameter("id");
        resp.setContentType("text/html; charset=UTF-8");
        PrintWriter w = resp.getWriter();
        if ((id == null) || (id.isBlank())) {
            w.println("<h2 style='text-align:center'>");
            w.println("参数错误</h2>");
            return;
        }
        try{
            Connection conn = DruidUtil.getConnection();
            PreparedStatement ps = conn.prepareStatement(
                "DELETE FROM student WHERE id = ?");
            ps.setInt(1, Integer.parseInt(id));
            int count = ps.executeUpdate();
            if (count > 0) {
                w.println("<h2  style='text-align:center'>");
                w.println("删除了 id=" + id + "的学生信息</h2>");
            }
            ps.close();
            conn.close();
        } catch (SQLException e){
            w.println(e.getMessage());
            log(e.getMessage());
```

            }
        }
    }

现在运行程序,在浏览器地址栏输入 StudentManDruid 的 URL 地址,将得到与图 7-2 所示一致的结果。

7.3 案例:将用户注册信息保存到数据库

用户注册是系统的基本功能。本节通过将用户注册信息保存到数据库的综合案例对本章的内容进行总结性应用。

7.3.1 案例目标

完善用户注册功能。编写一个 Java Web 程序,该程序接收用户在浏览器上输入的用户注册信息,除了包括用户名、密码、电话号码等基本信息外,还包括如下新的信息:用户的爱好和用户头像。当用户单击页面的"注册"按钮后,将用户注册信息发送到后台 Servlet 并保存在后台数据库中。用户信息保存到服务器后,返回一个页面,在该页面中用美观的方式显示用户的完整注册信息。

7.3.2 案例分析

这个程序首先应该包括一个注册页面,在这个页面上用户可以输入相关信息,当用户输入相关信息并单击提交按钮后,通过 POST 方法将注册信息提交服务端程序进行处理;服务端程序将数据保存到数据库中。成功保存用户数据后,将用户的完整信息再次通过一个页面显示在客户端。因此,这个程序包含如下内容。

(1) 数据库及其数据表设计:dbuser 数据库、user 表及 photo 表。因为允许一个用户上传多张头像,因此,需要设计 photo 表专门用于存储用户头像。

(2) 注册页面及其样式文件:index.html 和 index.css 文件。

(3) 服务端 Servlet 用于接收注册请求:Register.java,用于接收用户注册信息。

(4) 用户头像显示/下载服务端程序:Photo.java,一个 Servlet 程序,用于从数据库中读取用户头像并发送到客户端。

(5) POJO:Plain Old Java Object,User.java,用于描述用户对象。

(6) DAO:Data Access Object,UserDAO.java,用户操作数据表 user 表的信息。

(7) Druid 数据库连接池:DruidUtil.java,数据库连接池工具,使用在 7.2 节已经编写的代码即可。

应用程序的结构对应用程序的稳定性、编码效率、后期维护都有很大的影响。从上面的分析可以看出,所设计的程序具有如图 7-4 所示的分层结构。

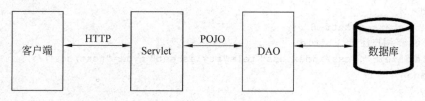

图 7-4　程序的分层结构

其中，register.html 及其布局构成了客户端层；Register 类及 Photo 类构成 Servlet 层；UserDAO 类构成了 DAO 层；User 类是 POJO，POJO 是 Servlet 层与 DAO 层进行数据交互的基本对象。采用这种结构，各个模块的功能定位明确清晰，便于编码和后期维护。

7.3.3　案例实施

首先进行数据库设计，包括两个方面内容：其一，设计 dbuser 数据库；其二，设计 user 数据表和 photo 数据表。创建数据库和数据表的 SQL 语句如下：

```sql
//创建 dbuser 数据库
CREATE SCHEMA 'dbuser' DEFAULT CHARACTER SET utf8mb4;
//创建 user 表
CREATE TABLE 'dbuser'.'user'(
  'id' INT NOT NULL AUTO_INCREMENT,
  'name' VARCHAR(45) NOT NULL,
  'password' VARCHAR(45) NOT NULL,
  'gender' VARCHAR(45) NOT NULL,
  'phone' VARCHAR(45) NOT NULL,
  'hobby' VARCHAR(200) NULL,
  PRIMARY KEY ('id'))
ENGINE=InnoDB
DEFAULT CHARACTER SET = utf8mb4;
//创建 photo 表
CREATE TABLE 'dbuser'.'photo'(
  'id' INT NOT NULL AUTO_INCREMENT,
  'user_id' INT NOT NULL,
  'image' LONGBLOB NOT NULL,
  PRIMARY KEY ('id'),
  INDEX 'user_id_idx' ('user_id' ASC) VISIBLE,
  CONSTRAINT 'user_id'
    FOREIGN KEY ('user_id')
    REFERENCES 'dbuser'.'user' ('id')
    ON DELETE NO ACTION
    ON UPDATE NO ACTION);
```

现在编写注册页面 index.html 及其样式文件 index.css。编写完成的 index.html 文件内容如下：

```html
<!DOCTYPE html>
<html lang="js">
```

```html
<head>
  <meta charset="utf-8">
  <title>用户注册</title>
  <link href="./css/index.css" rel="stylesheet" type="text/css" />
</head>
<body>
<h3>请填写注册信息</h3>
<div>
  <form enctype="multipart/form-data" method="POST" action="./Register">
    <label for="username">用户名</label>
    <input type="text" id="username" name="username" placeholder="用户名...">
    <label for="password1">密码</label>
    <input type="password" id="password1" name="password1" placeholder="密码...">
    <label for="password2">密码确认</label>
    <input type="password" id="password2" name="password2" placeholder="密码确认...">
    <label for="phone">电话号码</label>
    <input type="text" id="phone" name="phone" placeholder="电话号码...">
    <p>性别:
      <input type="radio" value="man" name="gender" checked/>男
      <input type="radio" value="woman" name="gender"/>女
    </p>
    <p>爱好:
      <input type="checkbox" value="reading" name="hobby" checked/>读书
      <input type="checkbox" value="running" name="hobby"/>跑步
      <input type="checkbox" value="chatting" name="hobby"/>聊天
      <input type="checkbox" value="traveling" name="hobby"/>旅游
    </p>
    <label for="photo">头像</label>
    <input type="file" multiple id="photo" name="photo" placeholder="选择头像文件...">
    <input type="submit" value="提交">
  </form>
</div>
</body>
</html>
```

注册页面的样式文件 index.css 文件内容如下：

```css
input[type=text], input[type=password], input[type=file], p, select{
    width: 100%;
    padding: 12px 20px;
    margin: 8px 0;
    display: inline-block;
    border: 1px solid #ccc;
    border-radius: 4px;
    box-sizing: border-box;
}
input[type=submit]{
    width: 100%;
    background-color: #4CAF50;
    color: white;
    padding: 14px 20px;
    margin: 8px 0;
```

```css
    border: none;
    border-radius: 4px;
    cursor: pointer;
}
input[type=submit]:hover{
    background-color: #45a049;
}
div{
    border-radius: 5px;
    background-color: #f2f2f2;
    padding: 20px;
}
```

当用户在注册页面填写了相关信息并单击"提交"按钮后,页面信息将被提交到 Register 服务端程序进行处理。Register.java 的代码如下:

```java
package com.ttt.servlet;
import com.ttt.User;
import com.ttt.dao.UserDAO;
import jakarta.servlet.ServletException;
import jakarta.servlet.annotation.MultipartConfig;
import jakarta.servlet.annotation.WebServlet;
import jakarta.servlet.http.HttpServlet;
import jakarta.servlet.http.HttpServletRequest;
import jakarta.servlet.http.HttpServletResponse;
import jakarta.servlet.http.Part;
import java.io.ByteArrayOutputStream;
import java.io.IOException;
import java.io.InputStream;
import java.io.PrintWriter;
import java.util.Base64;
import java.util.Collection;
import java.util.LinkedList;
import java.util.List;
@MultipartConfig
@WebServlet("/Register")
public class Register extends HttpServlet{
    UserDAO udao = new UserDAO();
    @Override
    protected void doPost(HttpServletRequest req, HttpServletResponse resp)
      throws ServletException, IOException{
        req.setCharacterEncoding("utf-8");
        String username=req.getParameter("username");
        String password1=req.getParameter("password1");
        String password2=req.getParameter("password2");
        String phone=req.getParameter("phone");
        String gender=req.getParameter("gender");
        String[] hobby=req.getParameterValues("hobby");
        resp.setContentType("text/html; charset=utf-8");
        StringBuffer sb=new StringBuffer();
```

```java
        if (username.isBlank())
            sb.append("必须输入用户名<br>");
        else if (password1.isBlank())
            sb.append("必须输入密码<br>");
        else if (password2.isBlank())
            sb.append("必须输入密码<br>");
        else if (!password1.equals(password2))
            sb.append("两次输入的密码不一致<br>");
        else if (phone.isBlank())
            sb.append("必须输入电话号码<br>");
        else if (gender.isBlank())
            sb.append("必须选择性别<br>");
        else if (hobby.length == 0)
            sb.append("必须选择爱好<br>");
        if (sb.length() !=0) {
            resp.getWriter().println(sb);
            return;
        }
        StringBuilder h=new StringBuilder();
        for(int i=0; i<hobby.length; i++){
            h.append(hobby[i]);
            if (i < hobby.length-1) h.append(";");
        }
        List<byte[]> photos=new LinkedList<>();
        Collection<Part> parts=req.getParts();
        for(Part p : parts){
            if (p.getSubmittedFileName() == null)
                continue;
            InputStream is=p.getInputStream();
            ByteArrayOutputStream baos=new ByteArrayOutputStream();
            byte[] b=new byte[1024];
            while(is.read(b)>0) {
                baos.write(b);
            }
            b=baos.toByteArray();
            photos.add(b);
        }
         User u= new User(0, username, password1, phone, gender, h.toString(),
         photos);
         int id=udao.addUser(u);
         u.setId(id);
        displayUser(resp, u);
    }
    private void displayUser(HttpServletResponse resp, User u) throws IOException{
        String begin="""
                <body style='width:800px; margin:0 auto'>
                <h3>用户信息</h3>
                <div>
```

```
            """;
        String end="""
                </div>
                </body>
                """;
        PrintWriter p = resp.getWriter();
        p.println(begin);
        p.println("<p>id: " + u.getId() + "</p>");
        p.println("<p>name: " + u.getName() + "</p>");
        p.println("<p>password: " + u.getPassword() + "</p>");
        p.println("<p>phone: " + u.getPhone() + "</p>");
        p.println("<p>gender: " + u.getGender() + "</p>");
        p.println("<p>hobby: " + u.getHobby() + "</p>");
        for(byte[] b : u.getPhotos()) {
            p.println("<img style='margin: 5px; width: 23%; height: 23%;
                float: left' " + "src='data:image/png;base64," + Base64.getEncoder().
                encodeToString(b) +"'/>");
        }
        p.println(end);
    }
}
```

Register 程序的代码较长，但是逻辑比较简单。程序首先从 HTTP 请求中获取相关参数信息，并以参数信息为基础创建了 User 对象：一个简单的 POJO 对象。

```
User u=new User(0, username, password1, phone, gender, h.toString(), photos);
```

然后，使用如下语句：

```
int id=udao.addUser(u);
u.setId(id);
```

将 User 对象的数据保存到数据库 user 表中，之后，再通过 displayUser() 方法将用户信息在浏览器中显示出来。在 displayUser() 方法中，使用 Base64 对用户头像二进制数据编码为字符串，使用 HTML 的<img>标签将 Base64 解码为图像数据并显示出来：

```
for(byte[] b : u.getPhotos()) {
    p.println("<img style='margin: 5px; width: 23%; height: 23%; float: left' " + "
        src='data:image/png;base64," + Base64.getEncoder().encodeToString(b)
        +"'/>");
}
```

对于小分辨率的图像，采用 Base64 将图像编码为字符串，并采用 HTML 的<img>标签内置的显示 Base64 图像的能力，可以减少浏览器对服务端的请求次数。

User 只是一个简单的 POJO 类。User.java 代码如下：

```
package com.ttt.pojo;
import java.util.List;
public class User{
    private int id;
    private String name;
    private String password;
```

```java
        private String phone;
        private String gender;
        private String hobby;
        private List<byte[]> photos;
        public User(int id, String name, String password, String phone,
           String gender, String hobby, List<byte[]> photos) {
            this.id=id;
            this.name=name;
            this.password=password;
            this.phone=phone;
            this.gender=gender;
            this.hobby=hobby;
            this.photos=photos;
        }
        public int getId(){
            return id;
        }
        public void setId(int id){
            this.id=id;
        }
        public String getName(){
            return name;
        }
        public void setName(String name){
            this.name=name;
        }
        public String getPassword(){
            return password;
        }
        public void setPassword(String password){
            this.password=password;
        }
        public String getPhone(){
            return phone;
        }
        public void setPhone(String phone){
            this.phone=phone;
        }
        public String getGender(){
            return gender;
        }
        public void setGender(String gender){
            this.gender=gender;
        }
        public String getHobby(){
            return hobby;
        }
        public void setHobby(String hobby){
```

```
        this.hobby=hobby;
    }
    public List<byte[]> getPhotos(){
      return photos;
    }
    public void setPhotos(List<byte[]> photos){
        this.photos=photos;
    }
}
```

现在看看 UserDAO 类。User DAO 类主要完成将用户注册数据保存到数据表 user 中。UserDAO.java 代码如下：

```
package com.ttt.dao;
import com.ttt.User;
import com.ttt.utils.DruidUtil;
import java.sql.Connection;
import java.sql.PreparedStatement;
import java.sql.ResultSet;
import java.sql.SQLException;
import java.util.List;
public class UserDAO{
    public int addUser(User u){
        Connection conn=DruidUtil.getConnection();
        int user_id;
        try{
            PreparedStatement ps1=conn.prepareStatement("INSERT INTO user " +
                "(name, password, gender, phone, hobby) VALUES (?, ?, ?, ?, ?)",
                PreparedStatement.RETURN_GENERATED_KEYS);
            PreparedStatement ps2=conn.prepareStatement("INSERT INTO photo " +
                "(user_id, image) VALUES (?, ?)");
            conn.setAutoCommit(false);
            ps1.setString(1, u.getName());
            ps1.setString(2, u.getPassword());
            ps1.setString(3, u.getGender());
            ps1.setString(4, u.getPhone());
            ps1.setString(5, u.getHobby());
            ps1.execute();
            ResultSet rs=ps1.getGeneratedKeys();
            if (rs.next())
                user_id=rs.getInt(1);
            else
                throw new RuntimeException("插入用户记录错误!");
            for(byte[] b : u.getPhotos()) {
                ps2.setInt(1, user_id);
                ps2.setBytes(2, b);
                ps2.execute();
            }
            conn.commit();
```

```
            ps1.close();
            ps2.close();
            conn.close();
        } catch (SQLException e) {
            try{
                conn.rollback();
            } catch (SQLException ex) {
                throw new RuntimeException(ex);
            }
            throw new RuntimeException(e);
        }
        return user_id;
    }
}
```

由于用户数据是存储在两张表中的：user 表用于存储用户的基本信息，photo 表用于存储用户的头像信息，因此，在 User DAO 类中开启了事务：

```
conn.setAutoCommit(false);
```

在事务开启之后，再将用户的基本信息和头像信息存储到相应的表中。如果两个信息均成功保存，则提交事务：

```
conn.commit();
```

否则，如果在操作数据表的过程中发生的任何异常，则回滚事务：

```
conn.rollback();
```

现在运行这个程序，将显示如图 7-5 所示的界面。

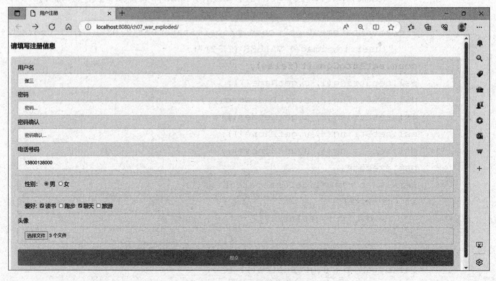

图 7-5　用户注册主页面

在其中填写用户信息，选择用户头像，然后单击"提交"按钮，提交数据到后台服务器，显示如图 7-6 所示的界面。

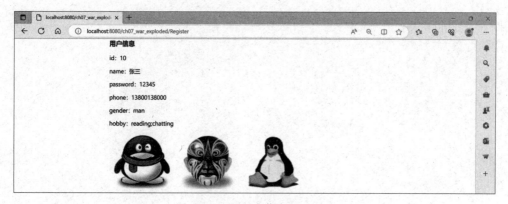

图7-6 注册用户信息显示页面

用户信息能够正确保存到数据库中,并在一个新页面中显示用户的完整信息。

7.4 练习:将图书信息保存到数据库

完善图书信息录入程序。要求首先制作一个图书录入页面,除了可以录入图书的基本信息,包括图书名称、出版社、编者、价格、图书介绍等基本信息外,还要求录入如下新的信息:图书的封面、图书的类型。当用户在页面上录入信息并单击"提交"按钮后,将图书信息保存在后台数据库中。图书信息保存到服务器后,返回一个页面,在该页面中,用美观的方式显示图书的完整信息。

第 2 部分

Java Web 表示技术

第 8 章　系统分层结构及 MVC 设计模式
第 9 章　JSP 表示技术
第 10 章　Thymeleaf 表示技术
第 11 章　JSON、JavaScript 和 Ajax

第 8 章　系统分层结构及 MVC 设计模式

具有一定规模的系统都是由多个功能部件协同工作进而达成系统总体功能目标的。在系统设计开发实践中,除了要达成系统的功能、性能、可用性目标外,还要考虑代码的可读性、未来可维护性等其他非功能及性能目标。这就要求在设计和开发系统时对系统要进行分层设计:采用面向对象的思想对功能部件进行封装,每个功能部件完成定义的功能、性能目标,并且尽量减少与部件之间的耦合,通过部件的协同达成系统整体目标。本章对系统分层设计技术和著名的 MVC 模式进行介绍。

8.1　程序功能部件之间的耦合度

设计软件系统的最为重要的原则是:程序功能部件之间尽量少地发生耦合。这里所谓的"耦合"是用来度量部件之间的关联关系密切程度的一个指标:如果两个组件之间仅仅是通过对象及其方法调用来完成协同的,则称这两个组件的"耦合度低";否则,如果两个组件之间除了通过对象及其方法调用完成协同外,还通过类似全局变量等方式才能完成部件之间的协同,则称两个组件的"耦合度高"。

部件之间较低的耦合度是系统设计所追求的目标。也就是说,在系统设计中应该尽量做到程序功能部件之间"解耦"。系统的分层设计是达成功能部件之间解耦的较好方式。

8.2　Java Web 程序的分层结构

分析一下典型的 Java Web 应用系统的运行过程:用户通过前端程序与系统交互,在前端程序中输入相应信息,然后将信息提交给后台的 Servlet 程序,后台 Servlet 接收用户信息,进行信息的合法性验证后,对信息进行进一步处理,也就是对用户的请求业务进行处理,之后,将数据写入数据库,并将处理结果数据发送给前端服务程序,前端服务程序将结果显示出来。这个过程可以用图 8-1 表示。

图 8-1　Java Web 程序的分层模式

(1) 前端页面：通过前端页面采集用户数据，并将数据发送到后台 Servlet 进行处理；在后台服务端程序处理完请求后，将处理结果发送给前端展示出来；针对前端浏览器，前端页面一般使用 HTML 技术、CSS 技术和 JavaScript 技术显示。

(2) 服务控制：由后台 Servlet 程序及页面生成程序构成，它一方面接收从前端页面发来的用户数据，并将数据交给业务处理部件进行处理，然后将处理结果以页面方式发送给前端，由前端将结果展示给用户。

(3) 业务处理：接收从服务控制发来的请求，进行业务处理，通过数据访问部件完成数据的访问和存储，最后将处理结果返回给服务控制部件。

(4) 数据访问：专门负责对数据库数据的访问，即 DAO(Data Access Object，数据访问对象)模式，包括对数据库数据的增、删、改、查操作。

(5) 数据存储：由数据库系统完成，是数据最终的存储点。

采用如图 8-1 所示的分层结构进行 Java Web 应用系统的设计，各个功能部件定位清晰，部件之间的耦合度较小，便于编码和实现，代码的可读性较好，后续的可维护性比较好。这种分层结构与 Java Web 设计界推崇的设计模式——MVC 设计模式非常契合。

8.3　Java Web 的 MVC 设计模式

设计模式是指在系统设计时遵循的一种理念、一种指导思想。MVC(Model-View-Controlller，模型-视图-控制器)是一种成功的软件设计思想。MVC 模式认为，一个系统由三个相互协同的部分构成，其中，Model 负责完成业务的处理；View 负责接收用户数据并将处理结果展示给用户；Controller 则负责协同 Model 和 View 工作，将特定的从 View 部分的业务请求交给适当的 Model 进行处理，并将处理结果交给适当的 View 进行展示。它们之间的关系如图 8-2 所示。

图 8-2　MVC 模式的逻辑过程

在 MVC 模式中，用户与 View 进行交互，View 将用户请求通过"服务请求"发给 Controller；Controller 分析用户请求，并根据分析结果"选择模型"，进而交给所选定的 Model 进行处理；Model 在进行服务处理过程中，从数据库读取数据，或者将数据存储到数据库中，之后，Model 将"处理结果"返回给 Controller；Controller 分析 Model 的处理结果，并"选择视图"对处理结果进行展示渲染；最后返回给 View，View 再将渲染结果展示给用户。

MVC 设计模式很好地表达了 Java Web 程序的处理逻辑。不要将这里所介绍的 MVC 模式与 8.2 节所介绍的"Java Web 程序的分层结构"搞混淆：MVC 是一种设计思想，而"Java Web 程序的分层结构"是 MVC 设计思想的具体实现方案，两者之间具有如图 8-3 所示的对应关系。

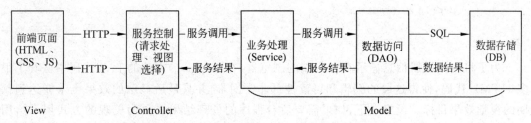

图 8-3　MVC 设计模式与系统分层结构之间的关系

因此,在 Java Web 的编程实践中,采用图 8-1 所示的方案对软件系统进行分层设计,是满足 MVC 设计模式思想的。实践证明,图 8-1 所示的 Java Web 系统的分层结果能够较好减少部件之间的耦合,是一种可靠的系统结构。

8.4　常用的 Java Web 表示技术

在介绍什么是"常用的 Java Web 表示技术"和具体有哪些"常用的 Java Web 表示技术"之前,回顾一下第 7 章的案例程序"7.3　案例:将用户注册信息保存到数据库"。在这个案例程序中,当用户填写注册信息并提交给服务端的 Register 程序后,Register 程序将用户的注册信息写入数据库,然后将用户的注册信息在客户端的浏览器中显示出来。为了显示用户一侧的信息,在 Register 程序中定义了 display()方法,其代码如下:

```java
private void displayUser(HttpServletResponse resp, User u) throws IOException
{
    String begin="""
        <body style='width:800px; margin:0 auto'>
        <h3>用户信息</h3>
        <div>
        """;
    String end="""
        </div>
        </body>
        """;
    PrintWriter p=resp.getWriter();
    p.println(begin);
    p.println("<p>id: " + u.getId() + "</p>");
    p.println("<p>name: " + u.getName() + "</p>");
    p.println("<p>password: " + u.getPassword() + "</p>");
    p.println("<p>phone: " + u.getPhone() + "</p>");
    p.println("<p>gender: " + u.getGender() + "</p>");
    p.println("<p>hobby: " + u.getHobby() + "</p>");
    for(byte[] b : u.getPhotos()) {
        p.println("<img style='margin: 5px; width: 23%; height: 23%; float: left' " + " src = 'data: image/png; base64," + Base64.getEncoder().encodeToString(b) +"'/>");
    }
```

```
            p.println(end);
        }
    }
```

为了将用户注册信息显示在客户端的浏览器,在 display() 方法的 Java 代码中嵌套了很多 HTML 代码,使得这段的编辑和可读性都非常困难,并且最后显示的效果也未能达到预期的视觉效果目标。设想一下,如果需要后台程序的处理结果以更为美观的方式展示给用户,势必要书写更多的 HTML 代码、CSS 代码、JavaScript 代码等,如果还是按照编写 display() 方法类似的方式编写页面代码,那将是难以想象的困难。因此,为了展示后台程序的处理结果,需要采用新的结果数据展示技术。这些将后台程序处理结果展示给用户的技术称为 Java Web 表示技术。常用的 Java Web 表示技术包括 JSP 表示技术和 Thymeleaf 表示技术。

第 9 章 JSP 表示技术

JSP 表示技术,也就是 Java Server Page 表示技术,是一种在服务端程序上生成浏览器页面的技术。简单来说就是将后台服务端程序的处理结果以动态方式生成 HTML 页面并在客户端浏览器上显示出来。JSP 技术是伴随 Servlet 技术的出现而出现的,是早期及现在主要的后台数据动态页面表示技术。

9.1 JSP 作为 MVC 的表示技术

在之前的案例程序中,为了将后台服务端程序的处理结果展示在客户端浏览器上,都是采用在 Servlet 代码中生成 HTML 页面的方式进行的。采用这种方式,在进行结果展示的 Servlet 代码中会存在 Java 代码与 HTML 代码、CSS 代码混杂的情况,导致程序的编写和可读性的巨大困难。为了解决这个问题,Java Web 提供了 JSP 技术。这种技术从另一个视角来审视这个问题:由于在进行结果展示的代码中,更多的是 HTML 代码、CSS 代码及 JavaScript 代码,因此,可以以一种特殊的方式将 Java 代码嵌入到 HTML 代码的方式来解决这个问题。为了对 JSP 技术有一个感性认识,先看一个简单的 JSP 程序的例子。

9.1.1 第一个 JSP 程序

新建一个名为 ch09 的 Java Web 工程。观察新建的 ch09 Java Web 工程,发现 IDEA 工具已经创建了一个名为 index.jsp 的文件。新建完成的工程如图 9-1 所示。

如图 9-1,仔细观察 index.jsp 代码,发现 index.jsp 的代码除了箭头 3 所指向的语句不是 HTML 语句外,其他语句都是(或者看起来是)HTML 标签语句。可以大胆猜测一下 index.jsp 程序的展示结果:在浏览器上显示 Hello World 文字,在这段文字下显示 Hello Servlet 的链接,单击这个链接,浏览器会访问服务端名称为 hello-servlet 的程序;URL 为 hello-servlet 的链接正是图 9-1 中箭头 1 所指向的 Servlet 程序的 URL 地址,如图 9-2 所示。

为了体现 JSP 程序"有点像" HTML 程序,但是又比 HTML 表达能力强,现在修改 index.jsp 为如下代码:

```
<%@ page contentType="text/html; charset=UTF-8" pageEncoding="UTF-8" %>
<!DOCTYPE html>
<html>
<head>
```

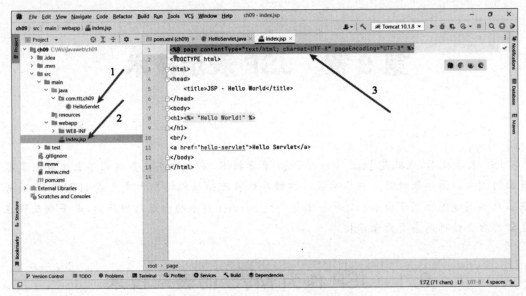

图 9-1 新建的 ch09 Java Web 工程

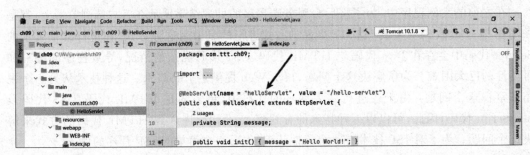

图 9-2 HelloServlet 的 URL 映射

```
    <title>JSP - Hello World</title>
</head>
<body>
<h1><%= "Hello World!" %></h1>
<br/>
<a href="hello-servlet">Hello Servlet</a>
<h2>
    <%= request.getContextPath()%>
</h2>
</body>
</html>
```

看看这句代码：

```
<h2>
    <%= request.getContextPath()%>
</h2>
```

其中的 request 就是表示 HTTP 的请求对象 HttpServletRequest，而 getContextPath() 方法

就是这个对象的一个方法。现在运行这个程序，显示如图 9-3 所示的结果。

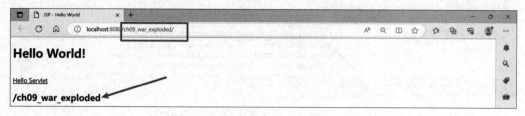

图 9-3　修改后的 index.jsp 显示结果

注意图 9-3 中箭头所指向的文字，它正是 ch09 工程的上下文路径。从图 9-3 的结果可以看出，在 JSP 代码中可以嵌入 Java 代码。确实是这样的，JSP 就是在 HTML 页面代码中嵌入了 Java 代码，因为 JSP 本质上就是 Servlet。为了理解这一点，需要对 JSP 的工作原理有所了解。

9.1.2　JSP 的工作原理

JSP 作为服务端程序，当用户访问一个 JSP 页面程序时，Servlet 容器（这里是 Tomcat 10）会把一个 JSP 页面程序自动转换为 Servlet 程序。浏览器请求 JSP 页面的过程如图 9-4 所示。

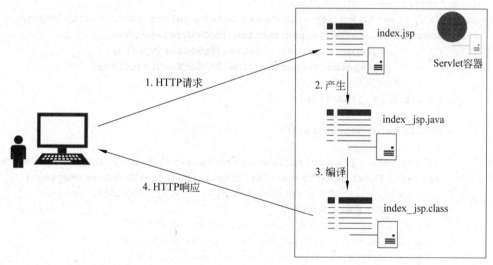

图 9-4　JSP 的工作过程

从图 9-4 可以看出，当客户端程序请求一个 JSP 页面时，Servlet 容器会将根据请求的 JSP 页面生成对应的 Servlet 程序，然后编译所生成的 Servlet 程序为其对应的字节码 class 程序，最后执行这个程序，并将这个程序的执行结果作为 HTTP 的响应发送到客户端。

可以直接观察到 Servlet 容器将 JSP 页面转换成 Servlet 程序的过程。对于 Tomcat 容器而言，可以在如图 9-5 所示的路径下找到 Tomcat 的工作目录。

这个目录下的子目录"\work\Catalina\localhost\ch09_war_exploded\org\apache\jsp"就是 Tomcat 生成和编译 ch09 工程 JSP 程序的结果目录，如图 9-6 所示。

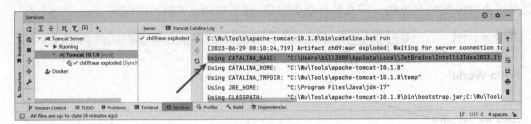

图 9-5 Tomcat 生成及编译 JSP 的工作目录

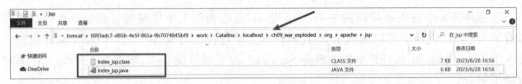

图 9-6 Tomcat 为 index.jsp 生成的中间代码

其中的 index_jsp.java 就是 Tomcat 为 index.jsp 生成的 Servlet 代码，index_jsp.class 就是编译后的字节码 class 程序。用文本编辑器打开 inde_jsp.java 源文件，省略其中对理解 JSP 不太重要的代码语句，index_isp.java 代码如下：

```
import jakarta.servlet.*;
import jakarta.servlet.http.*;
import jakarta.servlet.jsp.*;
public final class index_jsp extends org.apache.jasper.runtime.HttpJspBase
    implements org.apache.jasper.runtime.JspSourceDependent,
        org.apache.jasper.runtime.JspSourceImports,
        org.apache.jasper.runtime.JspSourceDirectives{
    ...    //此处省略了多行代码
    public void _jspInit(){
    }
    public void _jspDestroy(){
    }
    public void _jspService(final jakarta.servlet.http.HttpServletRequest
    request, final jakarta.servlet.http.HttpServletResponse response)
        throws java.io.IOException, jakarta.servlet.ServletException{
        ...    //此处省略了多行代码
        final jakarta.servlet.jsp.PageContext pageContext;
        jakarta.servlet.http.HttpSession session = null;
        final jakarta.servlet.ServletContext application;
        final jakarta.servlet.ServletConfig config;
        jakarta.servlet.jsp.JspWriter out = null;
        final java.lang.Object page = this;
        jakarta.servlet.jsp.JspWriter _jspx_out = null;
        jakarta.servlet.jsp.PageContext _jspx_page_context = null;
        try{
            ...    //此处省略了多行代码
            out.write("\n");
            out.write("<!DOCTYPE html>\n");
            out.write("<html>\n");
            out.write("<head>\n");
```

```
            out.write("<title>JSP - Hello World</title>\n");
            out.write("</head>\n");
            out.write("<body>\n");
            out.write("<h1>");
            out.print( "Hello World!" );
            out.write("\n");
            out.write("</h1>\n");
            out.write("<br/>\n");
            out.write("<a href=\"hello-servlet\">Hello Servlet</a>\n");
            out.write("<h2>\n");
            out.print( request.getContextPath());
            out.write("\n");
            out.write("</h2>\n");
            out.write("</body>\n");
            out.write("</html>\n");
        } catch (java.lang.Throwable t){
            ...    //此处省略了多行代码
        }
    }
}
```

在 Tomcat 对 index.jsp 生成的 Servlet 代码中,语句"public final class index_jsp extends org.apache.jasper.runtime.HttpJspBase"表示 index_jsp 是 HttpJspBase 的子类。但是,由于 HttpJspBase 是 HttpServlet 类的子类,所以,index_jsp 也是 HttpServlet 的子类,因此,index_jsp 就是 Servlet 程序。紧接着后面定义的两个方法:

```
public void _jspInit(){
}
public void _jspDestroy() {
}
```

这两个方法对应 Servlet 的 init()方法和 destroy()方法。现在重点关注一下_jspService()方法:本质上,_jspService()方法对应着 Servlet 的 service()方法。在_jspService()方法中,首先定义了一些对象变量,这里所定义的对象变量是在 JSP 代码可以直接使用的称为"内置对象"的对象。

注意语句"jakarta.servlet.jsp.JspWriter out = null;"所定义的 JspWriter 对象变量,这个变量对应 Servlet 的 PrintWriter 变量:向这个对象变量中输出的信息将直接发送到客户端服务程序中,如浏览器。再看看如下语句:

```
out.write("\n");
out.write("<!DOCTYPE html>\n");
out.write("<html>\n");
out.write("<head>\n");
out.write("<title>JSP - Hello World</title>\n");
out.write("</head>\n");
out.write("<body>\n");
out.write("<h1>");
out.print( "Hello World!" );
out.write("\n");
out.write("</h1>\n");
```

```
out.write("<br/>\n");
out.write("<ahref=\"hello-servlet\">Hello Servlet</a>\n");
out.write("<h2>\n");
out.print( request.getContextPath());
out.write("\n");
out.write("</h2>\n");
out.write("</body>\n");
out.write("</html>\n");
```

通过这一系列代码,将 JSP 程序中的 HTML 标签发送到客户端程序中。注意,在 index.jsp 代码中有以下语句:

```
<h2>
    <%= request.getContextPath()%>
</h2>
```

在对应的 index_jsp.java 中,Tomcat 将其生成如下语句:

```
out.write("<h2>\n");
out.print( request.getContextPath());
out.write("\n");
```

而其中的 request 对象正是 HttpServletRequest 类的对象。

从以上介绍的 JSP 程序的工作过程可以看出,JSP 程序本质上就是 Servlet,只是为了便于展示数据处理结果,将 Java 代码嵌入到 HTML 页面代码中而已。了解了 JSP 程序的本质,下面介绍如何编写 JSP 程序。

9.2 JSP 程序组成

从前面的介绍可以看出,JSP 程序"看起来像"一个 HTML 页面,但是可以在其中包含 Java 代码。总之,一个 JSP 页面程序由以下几部分组成。

(1) HTML 页面:HTML 标签、CSS、JavaScript 代码等。
(2) JSP 指令:常用的 JSP 指令包括 page 指令、taglib 指令。
(3) JSP 脚本:有三种脚本,也就是,声明脚本、表达式脚本、代码脚本。
(4) EL 表达式:通过"$"符号操作域对象中的数据。
(5) JSTL 标签库:一组 JSP 标签,可用于完成诸如基本输入/输出、流程控制、循环等操作。

9.2.1 JSP 指令

回顾一下本章开头创建 ch09 工程时 IDEA 自动生成的 index.jsp 代码,在 index.jsp 代码的第一句"<%@ page contentType="text/html; charset=UTF-8" pageEncoding="UTF-8" %>"就是 JSP 指令,这条指令称为 page 指令。JSP 常用的指令包括 page 指令和 taglib 指令。下面对 page 指令做介绍,关于 taglib 指令的含义及其使用将在 9.4 节介绍。

JSP 的 page 指令用于指定 JSP 页面的基本特征,这些基本特征包括页面的编码、页面的 MIME、是否生成 session 对象等。page 指令的一般格式如下：

<%@ page 属性名="属性值" 属性名="属性值" 属性名="属性值" …>

page 指令中可以使用的属性名及其含义如表 9-1 所示。

表 9-1 page 指令的属性名及其含义

序号	属性名	含 义
1	language	指定 JSP 页面所用的脚本语言，默认为 Java
2	import	类似于 Java 代码的 import 语句，导入 JSP 脚本需要的 Java 类库
3	session	是否为页面自动创建 HttpSession 对象，取值为 true 或者 fasle，默认值为 true，表示自动创建 session 对象，从而在 JSP 脚本中可以使用 session 对象维护会话
4	isErrorPage	指定页面是否为错误处理页面，取值可为 true 或者 false。默认值为 false。若为 true，表示该页面为错误处理页面，在 JSP 脚本代码中可以使用内置的 exception 对象来处理异常
5	errorPage	指向另一个 JSP 页面，表示当 JSP 页面程序产生异常时，则转到指定的页面进行处理。注意，所指向的页面的 isErrorPage 属性必须为 true
6	contentType	指定 JSP 页面的 MIME 类型和字符编码，相当于 Servlet 代码中的 HttpServletResponse 对象的 setContentType()方法
7	pageEncoding	指定 JSP 页面本身的编码方式

9.2.2 JSP 脚本

JSP 程序可以包含三种类型的脚本：声明脚本、表达式脚本、代码脚本。在此需要强调：在 JSP 程序中包含 JSP 脚本，会导致 JSP 程序的可读性和可维护性急剧降低，因此，不建议在 JSP 程序中使用 JSP 脚本，而应该使用后续介绍的 EL 表达式和 JSTL 标签库。在这里介绍 JSP 脚本的原因是为了兼顾维护以前使用了 JSP 脚本的代码需要。

1. 声明脚本

格式：

<%! java 代码 %>

作用：用于将 JSP 程序转换为 Servlet 程序时，在对应的 Servlet 类中定义脚本所指定的类变量或者成员方法。也就是说，所有的 JSP 声明脚本代码将被容器纳入 JSP 所对应的类中，相当于在对应的 Servlet 类中定义了类成员变量和成员方法。

例如：

```
<%!
    private String name;
    private List<String> levels;
    public int add(int a, int b){
        return a+b;
    }
%>
```

2. 表达式脚本

格式：

```
<%= 表达式 %>
```

作用：容器在将 JSP 程序转换为 Servlet 程序时，所有的 JSP 表达式脚本将被放置到对应 Servlet 类的_jspService()方法中，并且将每条 JSP 表达式脚本中的表达式解释为 out.print()方法的参数。

例如：

```
<%= 11+22 %>
<%= "Hello,World!" %>
<%= request.getContextPath() %>
```

3. 代码脚本

格式：

```
<% java 代码 %>
```

作用：容器在将 JSP 程序转换为 Servlet 程序时，所有的 JSP 代码脚本将被放置到对应 Servlet 类的_jspService()方法中。

例如：

```
<%
    for(int i=0; i<5; i++) {
        System.out.println("当前是" + i + "轮");
    }
%>
```

为了深入理解 JSP 脚本及观察 JSP 三种类型的脚本是如何被转换为 Servlet 代码的，下面举一个例子来观察一下 JSP 脚本的本质。为此，在 ch09 工程的 webapp 目录下新建 jsp 子目录，在这个子目录下新建名为 scripts.jsp 的程序。在 scripts.jsp 代码中，包括 JSP 的三种脚本代码。修改后的 scripts.jsp 代码如下：

```
<%@ page import="java.util.List" %>
<%@ page contentType="text/html;charset=UTF-8" language="java" %>
<html>
<head>
    <title>JSP 脚本示例</title>
</head>
<body>
    <h1>JSP 脚本示例</h1>
    <%!
        private String name;
        private List<String> levels;
        public int add(int a, int b){
            return a+b;
        }
    %>
    <p>荷塘的四面，远远近近，高高低低都是树，而杨柳最多。这些树将一片荷塘重重围住；只在小
```

路一旁，漏着几段空隙，像是特为月光留下的。</p>
<%= 11+22 %>
<%= "Hello,World!" %>
<%= request.getContextPath() %>
<p>忽然想起采莲的事情来了。采莲是江南的旧俗，似乎很早就有，而六朝时为盛；从诗歌里可以约略知道。</p>
<%
    for(int i=0; i<5; i++){
        System.out.println("当前是" + i + "轮");
    }
%>
</body>
</html>

运行这个程序，在浏览器地址栏输入 scripts.jsp，显示如图 9-7 所示的结果。

图 9-7　scripts.jsp 的运行结果

现在打开 scripts.jsp 对应的 Servlet 代码，如图 9-8 所示。

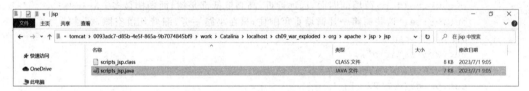

图 9-8　scripts.jsp 对应的 Servlet 文件

打开 scripts_jsp.java 文件，其内容如下：

```
...    //此处省略多行代码
public final class scripts_jsp extends org.apache.jasper.runtime.HttpJspBase
    implements org.apache.jasper.runtime.JspSourceDependent,
            org.apache.jasper.runtime.JspSourceImports,
            org.apache.jasper.runtime.JspSourceDirectives{
    private String name;
    private List<String> levels;
    public int add(int a, int b){
        return a+b;
    }
...    //此处省略多行代码
final jakarta.servlet.jsp.PageContext pageContext;
jakarta.servlet.http.HttpSession session = null;
final jakarta.servlet.ServletContext application;
final jakarta.servlet.ServletConfig config;
jakarta.servlet.jsp.JspWriter out = null;
```

```java
        final java.lang.Object page = this;
        jakarta.servlet.jsp.JspWriter _jspx_out = null;
        jakarta.servlet.jsp.PageContext _jspx_page_context = null;
        try{
          response.setContentType("text/html;charset=UTF-8");
          pageContext = _jspxFactory.getPageContext(this, request, response, null,
          true, 8192, true);
          _jspx_page_context = pageContext;
          application = pageContext.getServletContext();
          config = pageContext.getServletConfig();
          session = pageContext.getSession();
          out = pageContext.getOut();
          _jspx_out = out;
          out.write("\r\n");
          out.write("\r\n");
          out.write("<html>\r\n");
          out.write("<head>\r\n");
          out.write("    <title>JSP脚本示例</title>\r\n");
          out.write("</head>\r\n");
          out.write("<body>\r\n");
          out.write("    <h1>JSP脚本示例</h1>\r\n");
          out.write("\r\n");
          out.write("    ");
          out.write("\r\n");
          out.write("\r\n");
          out.write("<p>荷塘的四面,远远近近,高高低低都是树,而杨柳最多。\r\n");
          out.write("这些树将一片荷塘重重围住;只在小路一旁,漏着几段空隙,像是特为月光留下的。</p>\r\n");
          out.write("\r\n");
          out.write("    ");
          out.print( 11+22 );
          out.write("\r\n");
          out.write("    ");
          out.print( "Hello,World!" );
          out.write("\r\n");
          out.write("    ");
          out.print( request.getContextPath() );
          out.write("\r\n");
          out.write("\r\n");
          out.write("<p>忽然想起采莲的事情来了。采莲是江南的旧俗,似乎很早就有,而六朝时为盛;从诗歌里可以约略知道。</p>\r\n");
          out.write("\r\n");
          out.write("    ");
    for(int i=0; i<5; i++){
        System.out.println("当前是" + i + "轮");
    }
          out.write("\r\n");
          out.write("\r\n");
          out.write("\r\n");
          out.write("</body>\r\n");
          out.write("</html>\r\n");
```

```
        } catch (java.lang.Throwable t){
        ...    //此处省略多行代码
    }
}
```

scripts.jsp 代码对应的 Servlet 类为 scripts_jsp。注意 scripts_jsp 类中的代码：

```
private String name;
private List<String> levels;
public int add(int a, int b){
    return a+b;
}
```

这些代码就是从 JSP 的声明脚本转换而成的。注意，这些代码直接放置在 scripts_jsp 的类中，所定义的 name 变量和 add() 方法是 scripts_jsp 类的成员。再看看其中的代码：

```
out.print(11+22);
out.write("\r\n");
out.write("   ");
out.print("Hello,World!");
out.write("\r\n");
out.write("   ");
out.print(request.getContextPath());
```

这些代码是 JSP 的表达式标本对应的代码，它们直接放置在_jspService()方法中，并且将 JSP 表达式中的表达式值直接作为 out.print() 方法的参数。再看看其中的代码：

```
for(int i=0; i<5; i++) {
    System.out.println("当前是" + i + "轮");
}
```

这些代码是 JSP 的代码脚本对应的代码，它们直接放置在_jspService()方法中。

从这个例子可以清楚看到 JSP 脚本与其对应的 Servlet 类代码的关系。再次强调：JSP 脚本会导致 JSP 程序的可读性和可维护性急剧降低，因此，不建议在未来的 JSP 程序中嵌入 JSP 脚本，在展示请求处理结果时，应该使用即将介绍的 EL 表达式和 JSTL 标签库。

9.3 EL 表 达 式

EL 表达式结合 JSTL 标签库是编写 JSP 程序的最佳实践。EL 表达式是 JSP 的技术规范，其全称是表达式语言（expression language），它是一种简单的语言，提供了在 JSP 中简化表达式的方法，目的是尽量减少 JSP 页面中的 Java 代码，使得 JSP 页面的处理程序编写起来更加简洁，便于开发和维护。

9.3.1 EL 表达式基本语法及 EL 表达式内置对象

使用 EL 表达式的方式非常简单，就是通过"${属性名}"的方式来访问 JSP 程序数据。EL 表达式大括号中的属性名来自 JSP 程序的称为"EL 表达式隐式对象"的属性。EL 表达式共有 11 个内置的隐式对象,通过 EL 表达式的隐式对象,可以便利地操作程序的结果数据。EL 表达式的隐式对象的名称及其含义如表 9-2 所示。

表 9-2　EL 表达式的隐式对象的名称及其含义

序号	隐式对象名称	类　　别	类　　型	含　　义
1	pageContext	JSP 全局上下文对象	javax.servlet.PageContext	JSP 的 PageContext 对象,它表示整个 JSP 页面,可以获取或删除以下对象的任意属性:page(页面对象)、request(请求对象)、session(会话对象)、application(全局 Servlet 上下文)
2	pageScope	作用域对象	java.util.Map	取得 page 范围的属性名所对应的值
3	requestScope		java.util.Map	取得 request 范围的属性名所对应的值
4	sessionScope		java.util.Map	取得 session 范围的属性名所对应的值
5	applicationScope		java.util.Map	取得 application 范围的属性名所对应的值
6	param	请求参数对象	java.util.Map	相当于调用了 HttpServletRequest.getParameter(String name)方法,获得指定参数名称所对应的 String 类型值
7	paramValues		java.util.Map	相当于调用了 HttpServletRequest.getParameterValues(String name),获得指定参数名称所对应的 String[]类型值
8	header	请求头对象	java.util.Map	相当于调用了 HttpServletRequest.getHeader(String name),获得指定名称所对应的 String 类型的请求头值
9	headerValues		java.util.Map	相当于调用了 HttpServletRequest.getHeaders(String name),获得指定名称所对应的 String[]类型的请求头值
10	cookie	Cookie 对象	java.util.Map	相当于调用了 HttpServletRequest.getCookies(),获取 Cookie 的值
11	initParam	初始化参数对象	java.util.Map	相当于调用了 HttpServletContext.getInitParameter(String name),获得指定名称所对应的初始化参数值

在 EL 表达式中,可以直接使用表 9-2 所列的所有隐式对象,例如,${requestScope.username}表示要获取 requestScope 这个隐式对象中的名称为 username 的属性的值。在 11 个隐式对象中,除了 pageContext 对象是 javax.servlet.PageContext 类的对象外,其他 10 个隐式对象都是 java.util.Map 类的对象。通过 pageContext 对象的指定属性名可以获得 JSP 所对应的 Servlet 的几个重要对象,如表 9-3 所示。

表 9-3　pageContext 对象的属性及其含义

序号	属　性　名	含　　义
1	request	获取 JSP 对应的 Servlet 的 HttpServletRequest 对象。通过这个对象,可获取 HTTP 请求对象的各个属性值。例如,在 HttpServletRequest 对象中有一个 getMethod()方法,因此,可以使用 EL 表达式 ${pageContext.request.method} 获得 HTTP 请求方法的名称

续表

序号	属性名	含义
2	response	获取 JSP 对应的 Servlet 的 HttpServletResponse 对象。通过这个对象，可获取 HTTP 响应对象的各个属性值。例如，在 HttpServletResponse 对象中有一个 getContentType()方法，因此，可以使用 EL 表达式 ${pageContext.response.contentType}获得 HTTP 响应的内容类型
3	session	获取 JSP 对应的 Servlet 的 HttpSession 对象。通过这个对象，可获取 HttpSession 对象的各个属性值
4	servletContext	获取 JSP 对应的 Servlet 的 ServletContext 对象。通过这个对象，可获取 ServletContext 对象的各个属性值

下面举例说明 EL 表达式及其内置对象的使用。在这个例子中，通过浏览器请求 HelloEL 服务端 Servlet，并传递一个名称为 time 的参数。在 HelloEL 中，分别向 HttpServletRequest 对象等放置一些属性数据，再在 helloel.jsp 中通过 EL 表达式读取并显示在浏览器中。HelloEL.java 的代码如下：

```
package com.ttt.servlet;
import jakarta.servlet.RequestDispatcher;
import jakarta.servlet.ServletContext;
import jakarta.servlet.ServletException;
import jakarta.servlet.annotation.WebServlet;
import jakarta.servlet.http.HttpServlet;
import jakarta.servlet.http.HttpServletRequest;
import jakarta.servlet.http.HttpServletResponse;
import jakarta.servlet.http.HttpSession;
import java.io.IOException;
@WebServlet("/HelloEL")
public class HelloEL extends HttpServlet {
    @Override
    protected void doGet(HttpServletRequest req, HttpServletResponse resp)
      throws ServletException, IOException {
        req.setCharacterEncoding("UTF-8");
        req.setAttribute("school", "广州小学");
        HttpSession session = req.getSession(true);
        session.setAttribute("addr", "广州市");
        ServletContext context = req.getServletContext();
        context.setAttribute("age", 85);
        RequestDispatcher dispatcher=req.getRequestDispatcher("./jsp/helloel.jsp");
        dispatcher.forward(req, resp);
    }
}
```

在这个程序中，先通过如下代码：

```
req.setAttribute("school", "广州小学");
HttpSession session=req.getSession(true);
session.setAttribute("addr", "广州市");
ServletContext context=req.getServletContext();
context.setAttribute("age", 85);
```

分别向 HttpServletRequest 对象、HttpSession 对象和 ServletContext 对象中放置了名称为 school、addr 和 age 的属性，然后通过如下语句：

```
RequestDispatcher dispatcher=req.getRequestDispatcher("./jsp/helloel.jsp");
dispatcher.forward(req, resp);
```

将请求转发给 helloel.jsp 代码进行处理。helloel.jsp 的代码如下：

```
<%@ page contentType="text/html;charset=UTF-8" language="java" %>
<html>
<head>
    <title>EL 表达式及其内置对象</title>
    <meta charset="utf-8">
    <link href="${pageContext.request.contextPath}/css/style.css"
        rel="stylesheet" type="text/css" />
</head>
<body>
  <h2>方式一</h2>
  <p>${param.time}</p>
  <p>${requestScope.school}</p>
  <p>${sessionScope.addr}</p>
  <p>${applicationScope.age}</p>
  <h2>方式二</h2>
  <p>${param.time}</p>
  <p>${school}</p>
  <p>${addr}</p>
  <p>${age}</p>
</body>
</html>
```

在 helloel.jsp 代码中，通过 EL 表达式语句：

```
<p>${param.time}</p>
<p>${requestScope.school}</p>
<p>${sessionScope.addr}</p>
<p>${applicationScope.age}</p>
```

明确指明从哪个隐式对象获取属性信息。不仅如此，EL 表达式还提供了灵活的方式来获取属性信息：不用明确指明从哪个隐式对象获取信息。在没有明确指明从哪个属性获取属性信息的情况下，EL 表达式会按从小到大的作用域顺序获取属性信息，这个"从小到大"的作用域顺序是 pageScope、requestScope、sessionScope、applicationScope，也就是说，EL 表达式会先从 pageScope 对象中获取指定属性信息。若 pageScope 中不存在指定属性信息，则从 requestScope 中获取；若 requestScope 也不存在指定属性，则从 sessionScope 对象获取；若 sessionScope 对象还是不存在指定属性，则从 applicaitionScope 对象中获取；最后，若根本就存在指定属性，则 EL 表达式返回空串。因此，如下代码虽然没有指明隐式对象，但是是完全可以正常工作的：

```
<p>${param.time}</p>
<p>${school}</p>
<p>${addr}</p>
<p>${age}</p>
```

在 helloel.jsp 中使用了如下语句引入样式代码：

```
<link href="${pageContext.request.contextPath}/css/style.css"
    rel="stylesheet" type="text/css" />
```

样式文件 style.css 的代码如下：

```
body{
    width:100%;
    margin: 0 auto;
    text-align: center;
}
p{
    font-size: 20px;
    font-family:楷体;
}
```

运行这个程序，在浏览器的地址栏输入"HelloEL？time＝2023-07-11"，显示如图 9-9 所示的结果。

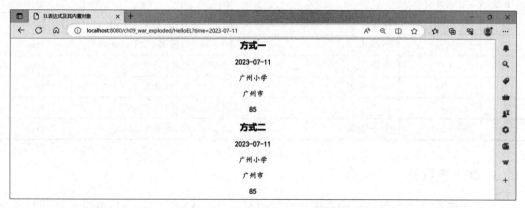

图 9-9　EL 表达式和隐式对象的基本使用

从逻辑上这个例子很简单：HelloEL 这个 Servlet 程序接收用户请求，进行适当的业务逻辑，然后将处理结果数据保存到 requestScope 或者 sessionScope 或者 applicationScope 中，最后通过 RequestDispatcher 将请求转发到某个 JSP 程序，最终将结果数据发送到客户端。这个例子虽然简单，但却是 Java Web 程序进行业务处理和结果数据展示的典型方式。这种方法本质上就是 MVC 编程模式。

9.3.2　EL 表达式运算符

使用 EL 表达式可以访问作用域数据，EL 还提供了一系列运算符用于数据操作，这些运算符包括算术运算符、关系运算符、逻辑运算符、条件运算符及 empty 运算符等。

1. 算术运算符

EL 的算术运算符可以进行各种算术运算，包括加减乘除等。EL 表达式的算术运算符及其含义如表 9-4 所示。

表 9-4 EL 的算术运算符及其含义

序号	运算符	含义	举例	结果
1	+	加法运算	${100+200}$	300
2	-	减法运算	${123.4-100}$	23.4
3	*	乘法运算	${12.3 * 200}$	2460
4	/,div	除法运算	${5/2}$,或者${5 div 2}$	2.5
5	%,mod	取余数运算	${5%2}$,或者${5 mod 2}$	1

2. 关系运算符

顾名思义,关系运算符就是判断两个数据的关系。EL 的关系运算符包括等于、不等于、大于、大于或等于、小于、小于或等于与等。EL 的关系运算符及其含义如表 9-5 所示。

表 9-5 EL 的关系运算符及其含义

序号	运算符	含义	举例	结果
1	==,eq	等于	${10 == 9}$,或者${10 eq 9}$	false
2	!=,ne	不等于	${10 != 9}$,或者${10 ne 9}$	true
3	<,lt	小于	${10<9}$,或者${10 lt 9}$	false
4	<=,le	小于或等于	${10 <= 9}$,或者${10 le 9}$	false
5	>,gt	大于	${10>9}$,或者${10 gt 9}$	true
6	>=,ge	大于或等于	${10 >= 9}$,或者${10 ge 9}$	true

3. 逻辑运算符

与其他编程语言类似,EL 的逻辑运算符包括 NOT、AND、OR。EL 逻辑运算符及其含义如表 9-6 所示。

表 9-6 EL 的逻辑运算符及其含义

序号	运算符	含义	举例	结果
1	&&,and	逻辑与,也就是"并且"	${(9==10) && (2==2)}$	false
2	||,or	逻辑或,也就是"或者"	${(9==10) && (2==2)}$	true
3	!,not	逻辑非,也就是"取反"	${!(9==10)}$	true

4. 条件运算符

条件运算符就是三元运算符,其一般形式是"条件？取值 1：取值 2"。也就是说,根据条件的真假,返回取值 1 或者取值 2 的值。例如,9==10? "A"："B",这个表达式最后返回字符串"B"。

5. empty 运算符

empty 运算符判断某个对象是否为空串,如果为空串则返回 true,否则返回 false。例如,${empty "123"}返回 false;${empty ""}返回 true;${empty null}返回 true。

EL 表达式结合下面即将介绍的 JSTL 标签库可以较好地发挥 JSP 程序的作用,因此,这里不再为 EL 表达式单独举例。

9.4 JSTL 标签及其使用

JSP 作为 MVC 模式的视图层,其主要任务是展示业务处理结果给前端用户。如前所述,JSP 本质上就是 Servlet,因此,可以在 JSP 中通过脚本嵌入 Java 代码。但是,如果在 JSP 代码中嵌入 Java 代码,会导致 HTML 代码和 Java 代码混在一起,导致代码的可读性和后期维护困难极大。因此,从 JSP 2.0 开始,引入了 JSTL 标签,从根本上解决了 JSP 代码中嵌入 Java 代码的弊端,结合 EL 表达式和 JSTL 标签库,使得 JSP 代码在很好地展示处理结果数据的同时,代码的可读性和可维护性得到极大的提升。

JSTL 是 javaserver pages standard tag library 的缩写,它提供了一组可以应用到 JSP 程序的标签,通过使用这组标签,可以完成包括基本输入/输出、流程控制、循环、XML 文件剖析、数据库查询及国际化和文字格式标准化等功能。JSTL 由五组不同功能的标签库组成,这五组标签库的名称及功能如表 9-7 所示。

表 9-7 五组标签库的名称及功能

序号	名 称	前缀	uri	备 注
1	核心标签库	c	jakarta.tags.core	使用频率高
2	格式化	fmt	jakarta.tags.fmt	较少使用
3	函数	fn	jakarta.tags.functions	较少使用
4	文档处理	xml	jakarta.tags.xml	不建议使用
5	数据库访问	sql	jakarta.tags.sql	不建议使用

下面对使用较多的核心标签库进行介绍。

9.4.1 如何使用 JSTL 标签库

为了在 JSP 程序中使用 JSTL 标签库,需要首先在程序工程的 pom.xml 文件中引入 JSTL 标签库的相关依赖。导入 JSTL 以来的语句如下:

```xml
<dependency>
    <groupId>org.glassfish.web</groupId>
    <artifactId>jakarta.servlet.jsp.jstl</artifactId>
    <version>3.0.1</version>
</dependency>
<dependency>
    <groupId>jakarta.servlet.jsp.jstl</groupId>
    <artifactId>jakarta.servlet.jsp.jstl-api</artifactId>
    <version>3.0.0</version>
</dependency>
```

在需要 JSTL 标签库的 JSP 代码的开头部分,使用如下语句导入需要使用的标签库。其中,语句:

<%@ taglib prefix="c" uri="jakarta.tags.core" %>

导入核心标签库。语句:

<%@ taglib prefix="fmt" uri="jakarta.tags.fmt" %>

导入格式化标签库。语句:

<%@ taglib prefix="fn" uri="jakarta.tags.functions" %>

导入函数标签库。语句:

<%@ taglib prefix="sql" uri="jakarta.tags.sql" %>

导入数据库处理标签库。语句:

<%@ taglib prefix="x" uri="jakarta.tags.xml" %>

导入文档处理标签库。

9.4.2 JSTL 核心标签

核心标签库是使用频率最高的 JSTL 标签库,使用核心标签库,可以完成输出数据、流程控制等功能。核心标签库的主要标签及其功能如表 9-8 所示。

表 9-8 核心标签库的主要标签及其功能

序号	标 签 名	功 能 描 述
1	<c:if>	构造类似 if…then 形式的条件表达式。当指定的条件为 true 时,则生成指定的结果数据到输出流
2	<c:choose> <c:when> <c:otherwise>	类似其他编程语言的 switch…case…default 语句,根据不同条件生成不同的结果数据到输出流
3	<c:forEach>	类似其他编程语言的 for 语句,完成对集合类数据的迭代操作
4	<c:forTokens>	根据指定的分隔符对数据进行迭代操作
5	<c:out>	将指定的数据输出到输出流中
6	<c:set>	把某个数据存入指定的变量中
7	<c:remove>	从指定的作用域中删除某个指定的数据
8	<c:url> <c:param>	集合<c:param>标签,构建带参数的 URL 地址
9	<c:redirect>	重定向到新的 URL
10	<c:import>	用来导入静态或动态文件内容,包括其他网站的文件

1. <c:if>标签及其使用

<c:if>标签用于简单的条件控制。<c:if>标签有三个属性,其中,test 属性是必需的返回结果为 true 或者 false 的测试条件;var 属性是用于保存条件测试结果的变量名;scope 属

性是 var 变量的作用域,有四个可选值：page、request、session 和 application。

下面编写一个简单的例子介绍<c:if>标签的使用。在这个例子中,用户在浏览器地址栏输入 JSTLEx,可以访问指定的 Servlet。在 Servlet 中,向 HttpServletRequest 对象、HttpSession 对象放置几个属性,然后使用 JSP 程序将这些信息显示在浏览器上。JSTLEx.java 代码如下：

```java
package com.ttt.servlet;
import jakarta.servlet.RequestDispatcher;
import jakarta.servlet.ServletException;
import jakarta.servlet.annotation.WebServlet;
import jakarta.servlet.http.HttpServlet;
import jakarta.servlet.http.HttpServletRequest;
import jakarta.servlet.http.HttpServletResponse;
import jakarta.servlet.http.HttpSession;
import java.io.IOException;
@WebServlet("/JSTLEx")
public class JSTLEx extends HttpServlet{
    @Override
    protected void doGet(HttpServletRequest req, HttpServletResponse resp)
      throws ServletException, IOException{
        req.setCharacterEncoding("UTF-8");
        req.setAttribute("name", "张三");
        String[] phones = {"13777777777", "13888888888", "13766666666"};
        req.setAttribute("phones", phones);
        HttpSession session = req.getSession(true);
        session.setAttribute("yah", "How do you do ?");
        session.setAttribute("tmp", 1234);
        RequestDispatcher dispatcher = req.getRequestDispatcher("./jsp/jstlif.jsp");
        dispatcher.forward(req, resp);
    }
}
```

这个 Servlet 代码比较简单,只是向 HttpServletRequest 对象、HttpSession 对象放置几个属性,然后将请求转发给 jstlif.jsp 程序将信息发送给前端客户。jstlif.jsp 的代码如下：

```jsp
<%@ page contentType="text/html;charset=UTF-8" language="java" %>
<%@ taglib prefix="c" uri="jakarta.tags.core" %>
<html>
<head>
  <title>JSTL 举例</title>
  <meta charset="utf-8">
  <link href="${pageContext.servletContext.contextPath}/css/style.css"
      rel="stylesheet" type="text/css" />
</head>
<body>
  <h2>JSTL 的 c:if 标签测试</h2>
  <c:if test="${requestScope.name == '张三'}" var="name1" scope="page">
    <p>你好,张三</p>
```

```
</c:if>
<p>c:if 标签中 name1 变量的值是：${pageScope.name1}</p>
<c:if test="${requestScope.name == '李四'}" var="name2" scope="page">
    <p>你好,李四</p>
</c:if>
<p>c:if 标签中 name2 变量的值是：${pageScope.name2}</p>
</body>
</html>
```

在 JSP 代码中,使用语句：

```
<c:if test="${requestScope.name == '张三'}" var="name1" scope="page">
    <p>你好,张三</p>
</c:if>
<p>c:if 标签中 name1 变量的值是：${pageScope.name1}</p>
```

测试 requestScope 对象中的 name 属性是否是"张三",如果是,则显示"你好,张三",并显示测试结果变量 name1 的值。运行这个程序,在浏览器地址栏输入 JSTLEx,显示如图 9-10 所示的结果。

图 9-10 <c:if>标签的使用

从 jstlif.jsp 代码中会发现,由于使用了 EL 表达式和 JSTL 标签,并且在 JSP 代码中并没有嵌入 Java 代码,使得 jstlif.jsp 代码的可读性较好,逻辑也比较清晰。这就是使用 EL 表达式和 JSTL 标签的优势。

JSTL 标签库的<c:if>标签观察

2. <c:choose>、<c:when>和<c:otherwise>标签及其使用

<c:choose>、<c:when>和<c:otherwise>标签可以完成对多条件的判断,并根据不同的结果输出不同的数据到输出流。<c:choose>标签没有属性,可以被认为是父标签,<c:when>、<c:otherwise>将作为其子标签来使用。<c:when>标签等价于 case 语句,它包含一个 test 属性,该属性表示需要判断的条件。<c:otherwise>标签没有属性,它等价于 default 语句。

下面举例说明<c:choose>、<c:when>和<c:otherwise>标签的使用。首先将 JSTLEx 程序中的语句：

```
RequestDispatcher dispatcher=req.getRequestDispatcher("./jsp/jstlif.jsp");
```

修改如下：

```
RequestDispatcher dispatcher=req.getRequestDispatcher("./jsp/jstlchoose.jsp");
```

然后,编写 jstlchoose.jsp 的代码如下：

```
<%@ page contentType="text/html;charset=UTF-8" language="java" %>
<%@ taglib prefix="c" uri="jakarta.tags.core" %>
<html>
<head>
  <title>JSTL举例</title>
  <meta charset="utf-8">
  <link href="${pageContext.servletContext.contextPath}/css/style.css"
        rel="stylesheet" type="text/css" />
</head>
<body>
  <h2>c:choose标签的使用举例</h2>
  <c:choose>
    <c:when test="${requestScope.name == '张三'}">
      <p>你好,张三</p>
    </c:when>
    <c:when test="${requestScope.name == '李四'}">
      <p>你好,李四</p>
    </c:when>
    <c:when test="${requestScope.name eq '王五'}">
      <p>你好,王五</p>
    </c:when>
    <c:otherwise>
      <p>不知道你是谁!</p>
    </c:otherwise>
  </c:choose>
</body>
</html>
```

在jstlchoose.jsp代码中,使用<c:choose>、<c:when>和<c:otherwise>标签对name属性的值进行判定,并根据判定结果输出不同的信息到客户端。运行这个程序,在浏览器地址栏输入JSTLEx,显示如图9-11所示的结果。

图9-11 <c:choose>标签使用示例

3. <c:forEach>标签及其使用

<c:forEach>标签类似其他编程语言的for语句,完成对集合类数据的迭代操作。<c:forEach>标签有多个属性,属性名及其含义如下。
- items:被循环迭代的变量,可选,无默认值。
- begin:起始序号,可选,默认值为0。
- end:结束序号,可选,默认值为最后元素的位置。
- step:步长,可选,默认值为1。
- var:迭代变量名,可选,无默认值。

- varStatus：迭代状态变量，可选，无默认值。它是 javax.servlet.jsp.jstl.core. LoopTagStatus 类型的变量。典型的属性中，index 表示当前位置索引，count 表示当前迭代元素计数，current 表示当前元素的引用。

使用<c:forEach>标签可以对 List 类型的变量、数组变量、Map 类型的变量等集合类型的变量进行迭代，逐个访问其中的每个元素并进行处理。下面举例说明<c:forEach>标签的使用。

首先将 JSTLEx 程序的中语句：

```
RequestDispatcher dispatcher=req.getRequestDispatcher("./jsp/jstlif.jsp");
```

修改如下：

```
RequestDispatcher dispatcher=req.getRequestDispatcher("./jsp/jstlforEach.jsp");
```

然后，编写 jstlforEach.jsp 的代码如下：

```
<%@ page contentType="text/html;charset=UTF-8" language="java" %>
<%@ taglib prefix="c" uri="jakarta.tags.core" %>
<html>
<head>
    <title>JSTL 举例</title>
    <meta charset="utf-8">
    <link href="${pageContext.servletContext.contextPath}/css/style.css"
        rel="stylesheet" type="text/css" />
</head>
<body>
    <h2>c:forEach 标签的使用举例</h2>
    <c:forEach items="${requestScope.phones}" var="phone" varStatus="status">
        Phone: ${status.index}, ${phone}, ${status.current}, ${status.count} <br>
    </c:forEach>
</body>
</html>
```

运行这个程序，在浏览器地址栏输入 JSTLEx，显示如图 9-12 所示的结果。

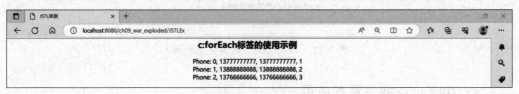

图 9-12 <c:forEach>标签的使用示例

JSTL 标签库的<c:forEach>标签观察

4. <c:forTokens>标签及其使用

<c:forTokens>标签通过指定分隔符将字符串分隔为一个数组并迭代它们。它与<c:forEach>具有相同的属性,但是多了一个 delims 属性,用于指定用于分割字符串的分隔符。下面举例说明<c:forTokens>标签的使用。首先将 JSTLEx 程序的中语句:

```
RequestDispatcher dispatcher=req.getRequestDispatcher("./jsp/jstlif.jsp");
```

修改如下:

```
RequestDispatcher dispatcher=req.getRequestDispatcher("./jsp/jstlforTokens.jsp");
```

然后,编写 jstlforTokens.jsp 的代码如下:

```
<%@ page contentType="text/html;charset=UTF-8" language="java" %>
<%@ taglib prefix="c" uri="jakarta.tags.core" %>
<html>
<head>
  <title>JSTL举例</title>
  <meta charset="utf-8">
  <link href="${pageContext.servletContext.contextPath}/css/style.css"
      rel="stylesheet" type="text/css" />
</head>
<body>
<h2>c:forTokens标签的使用举例</h2>
<c:forTokens items="${sessionScope.yah}" var="word" delims=" " varStatus=
"status">
   Phone: ${status.index}, ${word}, ${status.current}, ${status.count} <br>
</c:forTokens>
</body>
</html>
```

运行这个程序,在浏览器地址栏输入 JSTLEx,显示如图 9-13 所示的结果。

图 9-13 <c:forTokens>标签的使用举例

5. <c:out>、<c:set>和<c:remove>标签及其使用

<c:out>、<c:set>和<c:remove>标签分别用于输出指定的数据到客户端,设置指定变量的值和移除指定的变量。<c:out>标签的属性及其含义如下。

- value:要输出的值,必选,无默认值。
- default:当 value 的值不存在时,则输出 default 的值,可选,无默认值。
- escapeXML:是否忽略 XML 特殊字符,可选,默认值为 true。

<c:set>标签的属性及其含义如下。
- value：要存储的值，必选，无默认值。
- property：要修改的属性，可选，无默认值。
- target：要修改的属性所属的对象，可选，无默认值。
- var：要修改的变量的表变量名，可选，无默认值。
- scope：要修改的变量所属的作用域，可选，默认值为 page。

<c:remove>标签的属性及其含义如下。
- var：要移除的变量的名称，必选，无默认值。
- scope：要移除的变量所属的作用域，可选，默认值为 page。

下面举例说明<c:out>、<c:set>和<c:remove>标签的使用。首先将 JSTLEx 程序的中语句：

```
RequestDispatcher dispatcher=req.getRequestDispatcher("./jsp/jstlif.jsp");
```

修改如下：

```
RequestDispatcher dispatcher=req.getRequestDispatcher("./jsp/jstlout.jsp");
```

然后，编写 jstlout.jsp 的代码如下：

```
<%@ page contentType="text/html;charset=UTF-8" language="java" %>
<%@ taglib prefix="c" uri="jakarta.tags.core" %>
<html>
<head>
  <title>JSTL 举例</title>
  <meta charset="utf-8">
  <link href="${pageContext.servletContext.contextPath}/css/style.css"
      rel="stylesheet" type="text/css" />
</head>
<body>
  <h2>c:out c:set c:remove 标签的使用举例</h2>
  <c:out value="${requestScope.name}" escapeXml="true">
    name 的值为空
  </c:out>
  <br>
  <c:out value="${requestScope.name1}" escapeXml="true">
    name1 的值为空
  </c:out>
  <br>
  <c:out value="<h1>escapeXML 的值为 false</h1>" escapeXml="false"/>
  <br>
  <c:out value="<h1>escapeXML 的值为 true</h1>" escapeXml="true"/>
  <br>
  <c:set scope="request" var="newvar" value="12345"/>
  <c:out value="${newvar}" escapeXml="true">
    newvar 的值为空
  </c:out>
  <br>
  <c:remove var="newvar" scope="request"/>
```

```
    <c:out value="${newvar}" escapeXml="true">
      newvar 的值为空
    </c:out>
</body>
</html>
```

运行这个程序,在浏览器地址栏输入 JSTLEx,显示如图 9-14 所示的结果。

图 9-14 <c:out>、<c:set>、<c:remove>标签的使用举例

6. <c:url>和<c:param>标签及其使用

<c:url>标签将 URL 格式化为一个字符串,然后存储在一个变量中;<c:param>标签用来向包含或重定向的页面传递参数。<c:url>标签的属性如下。
- value:基础 URL,必选,无默认值。
- context:应用程序上下文,可选,默认值为当前应用程序的上下文。
- var:存储 URL 的变量名,必选,无默认值。
- scope:存储变量的作用域,可选,默认值为 page。

<c:param>标签的属性如下。
- name:参数名,必选,无默认值。
- value:参数值,可选,无默认值。

下面举例说明<c:url>和<c:param>标签的使用。首先将 JSTLEx 程序的中语句:

```
RequestDispatcher dispatcher=req.getRequestDispatcher("./jsp/jstlif.jsp");
```

修改如下:

```
RequestDispatcher dispatcher=req.getRequestDispatcher("./jsp/jstlurl.jsp");
```

然后,编写 jstlurl.jsp 的代码如下:

```
<%@ page contentType="text/html;charset=UTF-8" language="java" %>
<%@ taglib prefix="c" uri="jakarta.tags.core" %>
<html>
<head>
  <title>JSTL举例</title>
  <meta charset="utf-8">
  <link href="${pageContext.servletContext.contextPath}/css/style.css"
        rel="stylesheet" type="text/css" />
</head>
<body>
  <h2>c:url 和 c:param 标签的使用举例</h2>
```

```
    <c:url value="/HelloEL" context="${pageContext.servletContext.contextPath}"
      var="el">
      <c:param name="time" value="2023-07-13"/>
    </c:url>
    ${el}
    <br>
    <a href="${el}">访问 HelloEL</a>
  </body>
</html>
```

运行这个程序,在浏览器地址栏输入 JSTLEx,显示如图 9-15 所示的结果。

图 9-15　<c:url>和<c:param>标签的使用举例

在图 9-15 的界面中,单击"访问 HelloEL"链接,显示如图 9-16 所示的界面。

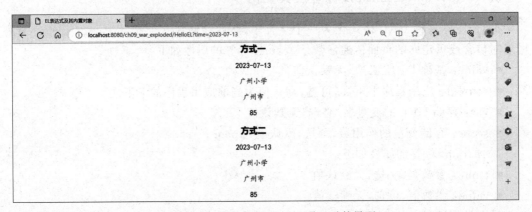

图 9-16　单击"访问 HelloEL"后显示的界面

7. <c:redirect>标签及其使用

<c:redirect>用于页面重定向,它的作用相当于调用了 JSP 对应的 HttpServletResponse 对象的 setRedirect 方法,这个标签可以配合<c:param>标签向目标链接传递请求参数。<c:redirect>标签的属性如下。

- url：重定向的目标 url,必选,无默认值。
- context：应用程序上下文,可选,默认值为当前应用程序的上下文。

下面举例说明<c:redirect>标签的使用。首先将 JSTLEx 程序的中语句:

```
RequestDispatcher dispatcher=req.getRequestDispatcher("./jsp/jstlif.jsp");
```

修改如下:

```
RequestDispatcher dispatcher=req.getRequestDispatcher("./jsp/jstlredirect.jsp");
```

然后,编写 jstlredirect.jsp 的代码如下:

```
<%@ page contentType="text/html;charset=UTF-8" language="java" %>
<%@ taglib prefix="c" uri="jakarta.tags.core" %>
<html>
<head>
  <title>JSTL举例</title>
  <meta charset="utf-8">
  <link href="${pageContext.servletContext.contextPath}/css/style.css"
        rel="stylesheet" type="text/css" />
</head>
<body>
  <h2>c:redirect 和 c:param 标签的使用举例</h2>
  <c:redirect url="/HelloEL" context="${pageContext.servletContext.
    contextPath}">
    <c:param name="time" value="2023-07-13"/>
  </c:redirect>
</body>
</html>
```

运行这个程序，在浏览器地址栏输入 JSTLEx，显示如图 9-17 所示的结果。

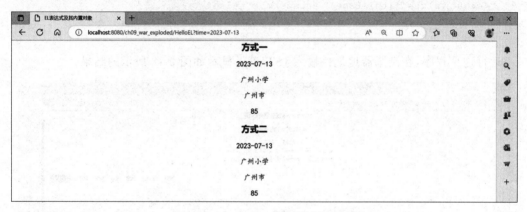

图 9-17　\<c:redirect\>标签用法示例

8. \<c:import\>标签及其使用

\<c:import\>标签用来导入静态或动态文件内容，包括来自其他网站的文件。\<c:import\>标签的属性如下。

- url：待导入资源的 URL，可以是本地资源，也可以是其他网站的资源，必选，无默认值。
- context：应用程序上下文，可选，默认值为当前应用程序的上下文。
- charEncoding：待导入资源的字符编码，可选，默认值为 ISO-8859-1。
- var：保存导入数据的变量名，可选，无默认值。
- scope：var 变量的作用域，可选，默认值为 page。
- varReader：用于提供 java.io.Reader 对象的变量，可选，无默认值。

下面举例说明\<c:import\>标签的使用。首先将 JSTLEx 程序的中语句：

```
RequestDispatcher dispatcher=req.getRequestDispatcher("./jsp/jstlif.jsp");
```

修改如下：

```
RequestDispatcher dispatcher=req.getRequestDispatcher("./jsp/jstlimport.jsp");
```

然后，编写 jstlimport.jsp 的代码如下：

```
<%@ page contentType="text/html;charset=UTF-8" language="java" %>
<%@ taglib prefix="c" uri="jakarta.tags.core" %>
<html>
<head>
  <title>JSTL 举例</title>
  <meta charset="utf-8">
  <link href="${pageContext.servletContext.contextPath}/css/style.css"
        rel="stylesheet" type="text/css" />
</head>
<body>
  <h2>c:import 标签的使用举例</h2>
  <c:import url="/jsp/jstlif.jsp" charEncoding="UTF-8"/>
  <hr>
  <c:import url="http://www.baidu.com" var="thisPage" charEncoding="UTF-8"/>
  <c:out value="${thisPage}" escapeXml="false" />
</body>
</html>
```

运行这个程序，在浏览器地址栏输入 JSTLEx，显示如图 9-18 所示的结果。

图 9-18　<c:import>标签的使用举例

9.5　JSP 最佳实践

一直以来，JSP 作为 Java Web 的数据表示方式之一得到了广泛应用。JSP 是灵活的，因为太灵活，所以难以驾驭。切记，只要将 JSP 作为数据结果展示技术，而不应该在其中嵌入 Java 代码进行业务处理,这种在 JSP 代码中嵌入 Java 代码是以往常见的方式,这种方式应该被摒弃。

结合 MVC 模式思想,开发 Java Web 程序的最佳实践是:客户端向后台服务程序发送请求,一般而言,后台服务程序是使用 Servlet 技术实现的。后台 Servlet 程序接收客户端的请求,获取相关请求参数,并根据请求信息将请求委托给业务服务程序,也就是 Service 程序进行处理。Service 程序根据是否需要操作数据库数据,创建 DAO 对象和 POJO 对象来完成对数据库数据的操作。一旦 DAO 完成对数据库数据的操作,将结果返回给 Service,Service 程序将结果进行适当处理后再返回给 Servlet 程序。Servlet 程序收到 Service 的返回结果后,将数据通过属性的形式保存到 HttpServletRequest 对象或者 HttpSession 对象或者 ServletContext 对象中,并将请求转发给目标 JSP 程序。目标 JSP 程序则从隐式对象中获取结果数据,结合 HTML、CSS、JavaScript,并使用 EL 表达式技术及 JSTL 技术将结果返回客户端。

9.6　案例:图书信息管理系统

作为对 Servlet 技术和 JSP 展示技术的综合使用,本案例程序设计一个图书信息管理系统,用于对图书信息进行管理。

9.6.1　案例目标

编写一个简单的图书管理系统,图书信息保存在数据库中。每本图书包含如下基本信息:名称、出版社、作者、图书简介、图书封面图片。在主页中,显示已经录入的所有图书信息,对已有的图书可以进行编辑,包括删除图书、修改图书信息,同时,在主页面中还包括一个"新增"按钮,单击该按钮,可以新增图书到数据库中。

9.6.2　案例分析

在可以对图书信息进行管理之前,需要首先登录到系统中,并且对于未登录的访问,系统将自动将访问引导到登录页面。因此,需要设计和编写如下程序文件。

(1) 登录页面:用 index.jsp 文件及其样式文件 index.css。

(2) 登录服务 Servlet:用 Login.java 文件,完成登录验证并创建 Session。登录成功后则直接跳转到主页面,登录失败则再次显示登录页面。

(3) 登录检查过滤器:用 MyFilter.java 文件,过滤器放行登录用户的访问,对未登录用户直接引导到登录页面。

(4) 数据 POJO:用 Book.java 文件,图书类是简单的 POJO 类。

(5) 主页面:用 books.jsp 文件及其样式文件 books.css 文件,图书信息主页面中显示所有已有图书信息。用户可以删除图书,修改图书信息和新增图书信息。

(6) 新增图书页面:用 addBook.jsp 文件,用户在这个页面可以录入新的图书信息。

(7) 图书管理 Servlet：用 BookServlet.java 文件，完成图书的增、删、改参数的接收，并将处理结果返回给客户端。

(8) 图书服务 Service：用 BookService.java 文件，完成具体的图书增、删、查、改服务，它被 BookServlet 使用，完成具体的业务处理。

(9) 图书数据 DAO：用 BookDAO.java 文件，完成具体数据信息的数据库操作。

(10) 数据库连接池：用 DruidUtil.java 文件及其配置文件 druid.properties，完成数据库连接池管理。

9.6.3 案例实施

首先创建数据库及相关数据表。在 MySQL 下创建 dbbooks 数据库和 book 数据表，相关语句如下：

```
CREATE SCHEMA 'dbbooks' DEFAULT CHARACTER SET utf8mb4;
CREATE TABLE 'dbbooks'.'book'(
  'id' INT NOT NULL AUTO_INCREMENT,
  'name' VARCHAR(45) NOT NULL,
  'publisher' VARCHAR(200) NOT NULL,
  'author' VARCHAR(45) NOT NULL,
  'memo' TEXT(2000) NULL,
  'cover' LONGBLOB NULL,
  PRIMARY KEY ('id'));
```

(1) 编写登录页面 index.jsp 及其样式文件 index.css。

```
<%@ page session="false" contentType="text/html; charset=UTF-8" pageEncoding="UTF-8" %>
<%@ taglib prefix="c" uri="jakarta.tags.core" %>
<!DOCTYPE html>
<html>
<head>
    <title>图书信息管理</title>
    <meta charset="utf-8">
    <link href="${pageContext.servletContext.contextPath}/css/index.css"
        rel="stylesheet" type="text/css" />
</head>
<body>
    <div>
        <h3>${requestScope.message}</h3>
        <form enctype="application/x-www-form-urlencoded" method="POST"
            target="_self" action="./Login">
            <label for="username">用户名</label>
            <input type="text" id="username" name="username" placeholder="用户名...">
            <label for="password">密码</label>
            <input type="password" id="password" name="password" placeholder="密码...">
            <input type="submit" value="提交">
        </form>
    </div>
```

```
</body>
</html>
```

样式文件 index.css 的代码如下：

```css
div{
    width: 600px;
    margin: 100px auto;
}
input[type=text], input[type=password]{
    width: 100%;
    padding: 12px 20px;
    margin: 8px 0;
    display: inline-block;
    border: 1px solid #ccc;
    border-radius: 4px;
    box-sizing: border-box;
}
input[type=submit]{
    width: 100%;
    background-color: #4CAF50;
    color: white;
    padding: 14px 20px;
    margin: 8px 0;
    border: none;
    border-radius: 4px;
    cursor: pointer;
}
input[type=submit]:hover{
    background-color: #45a049;
}
div{
    border-radius: 5px;
    background-color: #f2f2f2;
    padding: 20px;
}
```

（2）登录服务的 Servlet 并编写 Login.java 文件的代码。

```java
package com.ttt.servlet;
import com.ttt.pojo.Book;
import com.ttt.service.BookService;
import jakarta.servlet.RequestDispatcher;
import jakarta.servlet.ServletException;
import jakarta.servlet.annotation.WebServlet;
import jakarta.servlet.http.HttpServlet;
import jakarta.servlet.http.HttpServletRequest;
import jakarta.servlet.http.HttpServletResponse;
import jakarta.servlet.http.HttpSession;
import java.io.IOException;
import java.sql.SQLException;
import java.util.List;
@WebServlet("/Login")
```

```java
public class Login extends HttpServlet{
    @Override
    protected void doPost(HttpServletRequest req, HttpServletResponse resp)
            throws ServletException, IOException{
        req.setCharacterEncoding("UTF-8");
        String username = req.getParameter("username");
        String password = req.getParameter("password");
        if((username == null) || !(username.equalsIgnoreCase("admin")) ||
            (password == null) || !(password.equalsIgnoreCase("12345"))) {
            req.setAttribute("message", "登录失败！请重新登录");
            RequestDispatcher dispatcher = req.getRequestDispatcher("./index.
                jsp");
            dispatcher.forward(req, resp);
        }
        req.getSession();
        BookService bs = new BookService();
        try{
            List<Book> books = bs.getAllBooks();
            req.setAttribute("books", books);
            RequestDispatcher dispatcher = req.getRequestDispatcher("./jsp/
                books.jsp");
            dispatcher.forward(req, resp);
        } catch (SQLException e) {
            throw new RuntimeException(e);
        }
    }
}
```

（3）登录并检查过滤器并编写 MyFilter.java 文件的代码。

```java
package com.ttt.filter;
import jakarta.servlet.*;
import jakarta.servlet.annotation.WebFilter;
import jakarta.servlet.http.HttpServletRequest;
import jakarta.servlet.http.HttpServletResponse;
import jakarta.servlet.http.HttpSession;
import java.io.IOException;
@WebFilter(filterName = "MyFilter", urlPatterns = "/BookServlet")
public class MyFilter implements Filter{
    @Override
    public void doFilter(ServletRequest req, ServletResponse resp, FilterChain
        chain)
        throws IOException, ServletException{
        HttpServletRequest httpServletRequest = (HttpServletRequest)req;
        HttpSession session = httpServletRequest.getSession(false);
        if (session == null){
            String p = httpServletRequest.getContextPath();
            ((HttpServletResponse) resp).sendRedirect(p + "/index.jsp");
            return;
        }
        chain.doFilter(req, resp);
    }
```

}

(4) 编写数据 POJO 文件 Book.java 的代码。

```java
package com.ttt.pojo;
public class Book{
    private Integer id;
    private String name;
    private String publisher;
    private String author;
    private String memo;
    private byte[] cover;
    public Book(Integer id, String name, String publisher, String author,
                String memo, byte[] cover) {
        this.id = id;
        this.name = name;
        this.publisher = publisher;
        this.author = author;
        this.memo = memo;
        this.cover = cover;
    }
    public Integer getId(){
        return id;
    }
    public void setId(Integer id){
        this.id = id;
    }
    public String getName(){
        return name;
    }
    public void setName(String name){
        this.name = name;
    }
    public String getPublisher(){
        return publisher;
    }
    public void setPublisher(String publisher){
        this.publisher = publisher;
    }
    public String getAuthor(){
        return author;
    }
    public void setAuthor(String author){
        this.author = author;
    }
    public String getMemo(){
        return memo;
    }
    public void setMemo(String memo){
        this.memo = memo;
    }
    public byte[] getCover(){
```

```
        return cover;
    }
    public void setCover(byte[] cover){
        this.cover = cover;
    }
}
```

(5) 编写主页面对应的 books.jsp 文件及其样式文件 books.css 的代码。

books.jsp 的代码如下：

```
<%@ page session="false" contentType="text/html;charset=UTF-8" language=
"java" %>
<%@ taglib prefix="c" uri="jakarta.tags.core" %>
<html>
<head>
    <title>图书信息管理系统</title>
    <meta charset="utf-8">
    <link href="${pageContext.servletContext.contextPath}/css/books.css"
        rel="stylesheet" type="text/css" />
</head>
<body>
    <h2>图书信息管理系统</h2>
    <p><a href="${pageContext.servletContext.contextPath}/jsp/addBook.
       jsp">新增图书</a></p>
    <c:forEach items="${requestScope.books}" var="book">
        <div class="book">
            <img src="${pageContext.servletContext.contextPath}/
               BookServlet?f=cover&id=${book.id}" alt=""/>
            <p>书名：${book.name}</p>
            <p>出版社：${book.publisher}</p>
            <p>作者：${book.author}</p>
            <p>简介：${book.memo}</p>
            <p><a href="${pageContext.servletContext.contextPath}/
               BookServlet?f=delete&id=${book.id}">删除</a></p>
            <p><a href="${pageContext.servletContext.contextPath}/
               BookServlet?f=modify&id=${book.id}">修改</a></p>
        </div>
    </c:forEach>
</body>
</html>
```

books.css 的代码如下：

```
body{
    width: 1000px;
    margin: 0 auto;
}
h2{
    text-align: center;
}
.book{
    float: left;
```

```css
    width: 200px;
    height: 400px;
    border: 2px solid rgb(79, 185, 227);
    margin: 10px;
    padding: 10px;
}
img{
    display: block;
    width: 128px;
    height: 128px;
}
```

(6) 编写新增图书页面文件 addBook.jsp 及其样式文件 abook.css 的代码。

新增图书页面文件 addBook.jsp 的代码如下：

```jsp
<%@ page session="false" contentType="text/html;charset=UTF-8" language=
"java" %>
<%@ taglib prefix="c" uri="jakarta.tags.core" %>
<html>
<head>
    <title>图书信息管理系统</title>
    <meta charset="utf-8">
    <link href="${pageContext.servletContext.contextPath}/css/abook.css"
        rel="stylesheet" type="text/css" />
</head>
<body>
    <h2>新增/修改图书信息</h2>
    <div>
      <form enctype="multipart/form-data" method="POST"
            action="${pageContext.servletContext.contextPath}/BookServlet">
        <label for="id" hidden="hidden">id</label>
        <input type="text" id="id" name="id" value="${book.id}" style="display:
        none">
        <label for="name">图书名</label>
        <input type="text" id="name" name="name" value="${book.name}"
                                          placeholder="图书名称...">
        <label for="publisher">出版社</label>
        <input type="text" id="publisher" name="publisher"
                    value="${book.publisher}" placeholder="出版社...">
        <label for="author">作者</label>
        <input type="text" id="author" name="author"
                    value="${book.author}" placeholder="作者...">
        <label for="memo">图书简介</label>
        <input type="text" id="memo" name="memo"
                    value="${book.memo}" placeholder="图书简介...">
        <label for="cover">头像</label>
        <input type="file" id="cover" name="cover" placeholder="选择图书封面...">
        <input type="submit" value="提交">
      </form>
    </div>
</body>
```

```html
</html>
```

样式文件 abook.css 的代码如下：

```css
input[type=text], input[type=password], input[type=file], p, select{
    width: 100%;
    padding: 12px 20px;
    margin: 8px 0;
    display: inline-block;
    border: 1px solid #ccc;
    border-radius: 4px;
    box-sizing: border-box;
}
input[type=submit]{
    width: 100%;
    background-color: #4CAF50;
    color: white;
    padding: 14px 20px;
    margin: 8px 0;
    border: none;
    border-radius: 4px;
    cursor: pointer;
}
input[type=submit]:hover{
    background-color: #45a049;
}
div{
    border-radius: 5px;
    background-color: #f2f2f2;
    padding: 20px;
}
```

(7) 编写图书管理 Servlet 对应程序 BookServlet.java 的代码。

```java
package com.ttt.servlet;
import com.ttt.pojo.Book;
import com.ttt.service.BookService;
import jakarta.servlet.RequestDispatcher;
import jakarta.servlet.ServletException;
import jakarta.servlet.annotation.MultipartConfig;
import jakarta.servlet.annotation.WebServlet;
import jakarta.servlet.http.HttpServlet;
import jakarta.servlet.http.HttpServletRequest;
import jakarta.servlet.http.HttpServletResponse;
import jakarta.servlet.http.Part;
import java.io.ByteArrayOutputStream;
import java.io.IOException;
import java.io.InputStream;
import java.sql.SQLException;
import java.util.List;
@WebServlet("/BookServlet")
```

```java
@MultipartConfig
public class BookServlet extends HttpServlet{
    @Override
    protected void doGet(HttpServletRequest req, HttpServletResponse resp)
      throws ServletException, IOException{
        req.setCharacterEncoding("UTF-8");
        String f = req.getParameter("f");
        String id = req.getParameter("id");
        BookService bs = new BookService();
        switch(f){
            case "cover" ->{
                try{
                    byte[] cover = bs.getBookCoverById(Integer.parseInt(id));
                    resp.setContentType("image/jpeg");
                    resp.getOutputStream().write(cover);
                } catch (SQLException e){
                    throw new RuntimeException(e);
                }
            }
            case "delete" ->{
                try{
                    bs.deleteBookById(Integer.parseInt(id));
                    List<Book> books = bs.getAllBooks();
                    req.setAttribute("books", books);
                    RequestDispatcher dispatcher = req.getRequestDispatcher("./jsp/books.jsp");
                    dispatcher.forward(req, resp);
                } catch (SQLException e){
                    throw new RuntimeException(e);
                }
            }
            case "modify" ->{
                try{
                    Book b = bs.getBookById(Integer.parseInt(id));
                    req.setAttribute("book", b);
                    RequestDispatcher dispatcher = req.getRequestDispatcher("./jsp/addBook.jsp");
                    dispatcher.forward(req, resp);
                } catch (SQLException e){
                    throw new RuntimeException(e);
                }
            }
        }
    }
    @Override
    protected void doPost(HttpServletRequest req, HttpServletResponse resp)
      throws ServletException, IOException{
        req.setCharacterEncoding("UTF-8");
```

```java
            String id = req.getParameter("id");
            String name = req.getParameter("name");
            String publisher = req.getParameter("publisher");
            String author = req.getParameter("author");
            String memo = req.getParameter("memo");
            Part cover = req.getPart("cover");
            byte[] ci = null;
            if (cover != null){
                InputStream is = cover.getInputStream();
                ByteArrayOutputStream baos = new ByteArrayOutputStream();
                byte[] b = new byte[1024];
                while (is.read(b) > 0) {
                    baos.write(b);
                }
                ci = baos.toByteArray();
            }
            Book book = null;
            BookService bs = new BookService();
            if ((id != null) && !(id.isBlank())) {
                book = new Book(Integer.parseInt(id), name, publisher, author, memo, ci);
                try{
                    bs.updateBook(book);
                } catch (SQLException e){
                    throw new RuntimeException(e);
                }
            }
            else{
                book = new Book(-1, name, publisher, author, memo, ci);
                try{
                    bs.addBook(book);
                } catch (SQLException e){
                    throw new RuntimeException(e);
                }
            }
            List<Book> books = null;
            try{
                books = bs.getAllBooks();
            } catch (SQLException e){
                throw new RuntimeException(e);
            }
            req.setAttribute("books", books);
            RequestDispatcher dispatcher = req.getRequestDispatcher("./jsp/books.jsp");
            dispatcher.forward(req, resp);
    }
}
```

(8) 编写图书服务 Service 对应程序 BookService.java 的代码。

```java
package com.ttt.service;
```

```java
import com.ttt.dao.BookDAO;
import com.ttt.pojo.Book;
import java.sql.SQLException;
import java.util.List;
public class BookService{
    private finalBookDAO bdao;
    public BookService(){
        bdao = new BookDAO();
    }
    public List<Book> getAllBooks() throws SQLException{
        return bdao.findAllBooks();
    }
    public Book getBookById(int id) throws SQLException{
        return bdao.getBookById(id);
    }
    public byte[] getBookCoverById(int id) throws SQLException{
        return bdao.getBookCoverById(id);
    }
    public int deleteBookById(int id) throws SQLException{
        return bdao.deleteBookById(id);
    }
    public Book addBook(Book b) throws SQLException{
        return bdao.addBook(b);
    }
    public void updateBook(Book b) throws SQLException{
        bdao.updateBook(b);
    }
}
```

（9）编写图书数据 DAO 对应程序 BookDAO.java 的代码。

```java
package com.ttt.dao;
import com.ttt.pojo.Book;
import com.ttt.utils.DruidUtil;
import javax.swing.plaf.DimensionUIResource;
import java.sql.*;
import java.util.LinkedList;
import java.util.List;
public class BookDAO{
    public List<Book> findAllBooks() throws SQLException{
        Connection conn = DruidUtil.getConnection();
        LinkedList<Book> list = new LinkedList<>();
        Statement stat = conn.createStatement();
        ResultSet rs = stat.executeQuery("SELECT * FROM Book");
        while(rs.next()){
            Book b = new Book(rs.getInt("id"),
                rs.getString("name"), rs.getString("publisher"),
                rs.getString("author"), rs.getString("memo"), null);
            list.add(b);
        }
```

```java
        rs.close();
        stat.close();
        conn.close();
        return list;
    }
    public Book getBookById(int id) throws SQLException{
        Connection conn = DruidUtil.getConnection();
        PreparedStatement ps = conn.prepareStatement("SELECT * FROM book WHERE
        id =?");
        ps.setInt(1, id);
        ResultSet rs = ps.executeQuery();
        Book b = null;
        if (rs.next()){
            b = new Book(rs.getInt("id"), rs.getString("name"),
                rs.getString("publisher"), rs.getString("author"),
                rs.getString("memo"), rs.getBytes("cover"));
        }
        return b;
    }
    public byte[] getBookCoverById(int id) throws SQLException{
        Connection conn = DruidUtil.getConnection();
        PreparedStatement ps = conn.prepareStatement("SELECT cover FROM book
        WHERE id = ?");
        ps.setInt(1, id);
        ResultSet rs = ps.executeQuery();
        byte[] cover = null;
        if (rs.next()){
            cover = rs.getBytes("cover");
        }
        rs.close();
        ps.close();
        conn.close();
        return cover;
    }
    public int deleteBookById(int id) throws SQLException{
        Connection conn = DruidUtil.getConnection();
        PreparedStatement ps = conn.prepareStatement("DELETE FROM book WHERE
        id = ?");
        ps.setInt(1, id);
        int count = ps.executeUpdate();
        ps.close();
        conn.close();
        return count;
    }
    public Book addBook(Book b) throws SQLException{
        Connection conn = DruidUtil.getConnection();
        PreparedStatement ps = conn.prepareStatement(
            "INSERT INTO book(name,publisher,author,memo,cover) VALUES(?, ?, ?,
```

```java
                ?, ?)",
                PreparedStatement.RETURN_GENERATED_KEYS);
        ps.setString(1, b.getName());
        ps.setString(2, b.getPublisher());
        ps.setString(3, b.getAuthor());
        ps.setString(4, b.getMemo());
        ps.setBytes(5, b.getCover());
        if (ps.execute()){
            ResultSet rs = ps.getGeneratedKeys();
        if (rs.next())
            b.setId(rs.getInt(1));
        }
        return b;
    }
    public void updateBook(Book b) throws SQLException{
        Connection conn = DruidUtil.getConnection();
        if ((b.getCover() == null) || (b.getCover().length <= 0)) {
            PreparedStatement ps = conn.prepareStatement(
            "UPDATE book SET name =?," + "publisher =?, author = ?, memo = ? WHERE
            id = ?");
            ps.setString(1, b.getName());
            ps.setString(2, b.getPublisher());
            ps.setString(3, b.getAuthor());
            ps.setString(4, b.getMemo());
            ps.setInt(5, b.getId());
            ps.executeUpdate();
            ps.close();
            conn.close();
        }
        else{
            PreparedStatement ps = conn.prepareStatement(
            "UPDATE book SET name =?," + "publisher =?, author = ?, memo = ?,
            cover = ? WHERE id = ?");
            ps.setString(1, b.getName());
            ps.setString(2, b.getPublisher());
            ps.setString(3, b.getAuthor());
            ps.setString(4, b.getMemo());
            ps.setBytes(5, b.getCover());
            ps.setInt(6, b.getId());
            ps.executeUpdate();
            ps.close();
            conn.close();
        }
    }
}
```

（10）编写数据库连接池程序 DruidUtil.java 及其配置文件 druid.properties 的代码。

数据库连接池程序 DruidUtil.java 代码如下：

```java
package com.ttt.utils;
import com.alibaba.druid.pool.DruidDataSourceFactory;
import javax.sql.DataSource;
import java.io.InputStream;
import java.sql.Connection;
import java.sql.SQLException;
import java.util.Properties;
public class DruidUtil{
    private static DataSource ds = null;
    public static Connection getConnection(){
        if (ds == null){
            InputStream in = DruidUtil.class.getClassLoader().
                getResourceAsStream("druid.properties");
            Properties prop = new Properties();
            try{
                prop.load(in);
                ds = DruidDataSourceFactory.createDataSource(prop);
            } catch (Exception e){
                throw new RuntimeException(e);
            }
        }
        try{
            return ds.getConnection();
        } catch (SQLException e){
            throw new RuntimeException(e);
        }
    }
}
```

配置文件 druid.properties 代码如下：

```
driverClassName=com.mysql.cj.jdbc.Driver
url=jdbc:mysql://localhost:3306/dbbooks?serverTimezone=Asia/Shanghai
username=root
password=12345
initialSize=10
maxActive=50
minIdle=5
maxWait=6000
```

运行这个程序，显示如图 9-19 所示的登录页面。

在用户名和密码框分别输入 admin 和 12345，登录系统，显示如图 9-20 所示的界面。

在图 9-20 的界面中单击"新增图书"链接，显示如图 9-21 的界面，在这个界面可以录入新的图书信息。

录入多本图书后，显示如图 9-22 所示的界面。

在图 9-22 的界面中单击"删除"链接，可以删除指定图书；单击"修改"链接，可以修改图书信息。

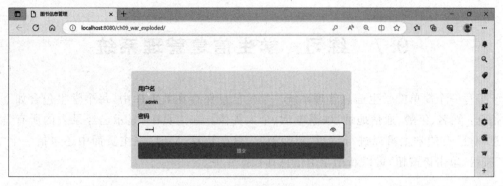

图 9-19　图书管理系统登录页面

图 9-20　图书管理系统主页面

图 9-21　图书信息录入界面

图 9-22　图书信息操作界面

9.7 练习：学生信息管理系统

编写一个简单的学生信息管理系统。学生信息保存在数据库中，每个学生包含如下基本信息：姓名、年龄、通信地址、联系电话、个人头像。在主页中，显示已经录入的所有学生信息，对已有的学生可以进行编辑，包括删除、修改信息，同时，在主页面中还包括一个"新增"按钮，单击该按钮，可以新增学生信息到数据库中。

第 10 章 Thymeleaf 表示技术

Thymeleaf 是一种服务端的模板引擎,是 Spring 框架推荐的前端表示框架。Thymeleaf 引擎可以将指定的模板与结果数据结合,形成期望的数据结果表示形式。Thymeleaf 可以处理的模板包括 HTML、XML、CSS、JavaScript、文本等。本章介绍如何使用 Thymeleaf 引擎处理 HTML 模板,并将 Java Web 程序的处理结果使用 Thymeleaf 模板进行展示。

10.1 Thymeleaf 作为 MVC 表示技术

在第 9 章介绍了如何使用 JSP 技术将 Java Web 程序的处理结果在客户端展示出来。本质上,Thymeleaf 在 Java Web 程序中的地位和角色等同于 JSP,也就是说,Thymeleaf 可以完全取代 JSP 在客户端展示 Java Web 处理结果数据。在详细介绍 Thymeleaf 的使用之前,先看一个简单的例子。要在 Java Web 程序中使用 Thymeleaf,需要先将 Thymeleaf 导入到项目中。

10.1.1 导入 Thymeleaf 到项目工程

新建一个名为 ch10 的 Java Web 工程并导入 Thymeleaf 相关依赖。截至编写本书时,Thymeleaf 的最新版本是 3.1.1 版。新建完成的项目并且导入 Thymeleaf 依赖的工程如图 10-1 所示。

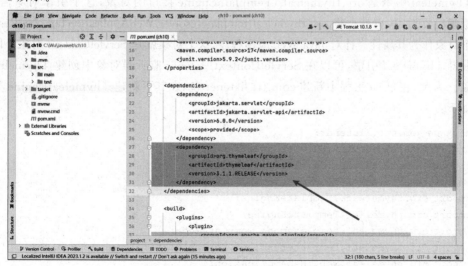

图 10-1 新建的 ch10 Java Web 工程

也就是说,需要在工程的 pom.xml 文件中增加以下依赖:

```xml
<dependency>
    <groupId>org.thymeleaf</groupId>
    <artifactId>thymeleaf</artifactId>
    <version>3.1.1.RELEASE</version>
</dependency>
```

10.1.2 创建 Thymeleaf 引擎

Thyemeleaf 是一个模板引擎,针对 HTML 模板而言,它将 HTML 模板中的引用替换为真实的结果数据。Thymeleaf 引擎的工作原理如图 10-2 所示。

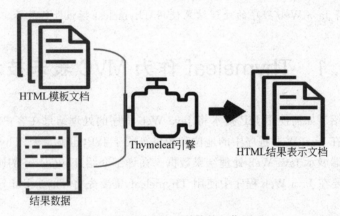

图 10-2 Thymeleaf 引擎的工作原理

从图 10-2 可以看出,Thymeleaf 引擎会扫描 HTML 模板文档,将模板中的数据引用替换为结果数据中的真实数据,然后将生成的结果文档发送到客户端,进而在客户端浏览器上显示出来。因此,在可以使用 Thymeleaf 引擎之前,需要首先创建 Thymeleaf 引擎对象。

Thymeleaf 引擎是 org.thymeleaf.ITemplateEngine 接口的对象,这个引擎对象需要一个模板解析器对象,也就是将 org.thymeleaf.templateresolver.WebApplicationTemplateResolver 的一个对象作为其属性。针对 Java Web 应用程序,这个对象就是 ServletContext 对象。因此,为了获得 Thymeleaf 引擎,可以在 ServletContextListener 监听器对象中创建 Thymeleaf 引擎对象。为此,在 ch10 工程中新建 com.ttt.listener 包,在其下新建 ThymeleafEngine 类,其代码如下:

```java
package com.ttt.listener;
import jakarta.servlet.ServletContextEvent;
import jakarta.servlet.ServletContextListener;
import jakarta.servlet.annotation.WebListener;
import org.thymeleaf.ITemplateEngine;
import org.thymeleaf.TemplateEngine;
import org.thymeleaf.templatemode.TemplateMode;
import org.thymeleaf.templateresolver.WebApplicationTemplateResolver;
import org.thymeleaf.web.IWebApplication;
import org.thymeleaf.web.servlet.JakartaServletWebApplication;
```

```java
@WebListener
public class ThymeleafEngine implements ServletContextListener {
    public static final String THYMELEAF_ENGINE = "com.ttt.thymeleaf.engine";
    @Override
    public void contextInitialized(ServletContextEvent sce) {
        JakartaServletWebApplication application = JakartaServletWebApplication
                .buildApplication(sce.getServletContext());
        ITemplateEngine templateEngine = getTemplateEngine(application);
        sce.getServletContext().setAttribute(THYMELEAF_ENGINE, templateEngine);
    }
    private ITemplateEngine getTemplateEngine(IWebApplication application) {
        TemplateEngine templateEngine = new TemplateEngine();
        WebApplicationTemplateResolver templateResolver =
            getTemplateResolver(application);
        templateEngine.setTemplateResolver(templateResolver);
        return templateEngine;
    }
    privateWebApplicationTemplateResolver getTemplateResolver(
            IWebApplication application) {
        WebApplicationTemplateResolver templateResolver =
            new WebApplicationTemplateResolver(application);
        //HTML 是默认的模板,这里明确设置是为了让代码更好懂
        templateResolver.setTemplateMode(TemplateMode.HTML);
        //配置字符集,避免中文乱码
        templateResolver.setCharacterEncoding("UTF-8");
        //这个设置将使得类似 home 映射为"/WEB-INF/templates/home.html"
        templateResolver.setPrefix("/WEB-INF/templates/");
        templateResolver.setSuffix(".html");
        //设置模板的 TTL 为 1 小时
        templateResolver.setCacheTTLMs(3600000L);
        //Cache 的默认值为 true。如果希望模板被更改时自动更新,设置这个值为 false
        templateResolver.setCacheable(true);
        return templateResolver;
    }
    @Override
    public void contextDestroyed(ServletContextEvent sce) {
    }
}
```

因为 ThymeleafEngine 是 ServletContextListener 监听器,在 Tomcat 加载该 Java Web 应用程序时会调用其 contextInitialized()方法。在这个方法中,通过如下语句:

```java
JakartaServletWebApplication application = JakartaServletWebApplication
        .buildApplication(sce.getServletContext());
ITemplateEngine templateEngine = getTemplateEngine(application);
sce.getServletContext().setAttribute(THYMELEAF_ENGINE, templateEngine);
```

创建了 ITemplateEngine 对象,并将它作为名称为 THYMELEAF_ENGINE 的属性放置到 ServletContext 对象中,在需要 Thymeleaf 引擎时,直接通过属性 THYMELEAF_ENGINE 即可获得 Thymeleaf 引擎对象的引用。

WebApplicationTemplateResolver 的对象是 Thymeleaf 引擎的重要属性,通过这个对

象,可以配置Thymeleaf引擎的模板类型、模板前缀、模板后缀等相关参数。

程序中粗体字的配置代码要求所有的Thymeleaf模板代码放置在"/WEB-INF/templates/"路径下,并且必须是HTML模板。现在已经创建了Thymeleaf引擎对象,并且配置了Thymeleaf引擎的参数,可以在程序中使用该Thymeleaf引擎了。

10.1.3 使用Thymeleaf引擎生成结果页面

现在编写一个简单的Servlet程序——FirstThymeleaf。在这个Servlet程序中,首先获取Thymeleaf引擎,并在Context对象中放置一个属性名为name的数据,然后通过Thymeleaf引擎将数据显示到home.html的模板文档中。FirstThymeleaf.java的代码如下。

```java
package com.ttt.servlet;
import jakarta.servlet.ServletException;
import jakarta.servlet.annotation.WebServlet;
import jakarta.servlet.http.HttpServlet;
import jakarta.servlet.http.HttpServletRequest;
import jakarta.servlet.http.HttpServletResponse;
import java.io.IOException;
import org.thymeleaf.TemplateEngine;
import org.thymeleaf.context.WebContext;
import org.thymeleaf.web.IWebExchange;
import org.thymeleaf.web.servlet.JakartaServletWebApplication;
import com.ttt.listener.ThymeleafEngine;
@WebServlet("/first")
public class FirstThymeleaf extends HttpServlet{
    @Override
    protected void doGet(HttpServletRequest req, HttpServletResponse resp)
      throws ServletException, IOException{
        TemplateEngine templateEngine = (TemplateEngine) getServletContext().
            getAttribute(ThymeleafEngine.THYMELEAF_ENGINE);
        IWebExchange webExchange = JakartaServletWebApplication
            .buildApplication(getServletContext())
            .buildExchange(req, resp);
        WebContext context = new WebContext(webExchange);
        context.setVariable("name", "Thymeleaf使用举例");
        //req.setAttribute("name", "Thymeleaf使用举例");
        //IWebSession session = webExchange.getSession();
        //session.setAttributeValue("name", "Thymeleaf使用举例");
        resp.setContentType("text/html; charset=UTF-8");
        templateEngine.process("home", context, resp.getWriter());
    }
}
```

在这个Servlet程序中,首先使用以下语句:

```
TemplateEngine templateEngine = (TemplateEngine) getServletContext().getAttribute
(ThymeleafEngine.THYMELEAF_ENGINE);
```

以上语句获得 Thymeleaf 引擎对象,这个 Thymeleaf 引擎对象是在 Tomcat 启动该 Java Web 程序时,在 ServletContextListener 监听器中创建的。然后使用以下语句:

```
IWebExchange webExchange = JakartaServletWebApplication
    .buildApplication(getServletContext()).buildExchange(req, resp);
WebContext context = new WebContext(webExchange);
```

以上语句构建了 Thymeleaf 引擎的 WebContext 对象,通过这个对象可将应用程序的上下文数据传递到 Thymeleaf 引擎中,以便 Thymeleaf 引擎将结果数据放置到 HTML 模板中。之后,再使用以下语句:

```
context.setVariable("name", "Thymeleaf 使用举例");
//req.setAttribute("name", "Thymeleaf 使用举例");
//IWebSession session = webExchange.getSession();
//session.setAttributeValue("name", "Thymeleaf 使用举例");
```

以上语句将名称为 name 的属性放置到 Thymeleaf 的 context 对象中。注意,可以根据程序需要,将数据放置到 req 对象或者 session 对象中。然后使用以下语句:

```
resp.setContentType("text/html; charset=UTF-8");
templateEngine.process("home", context, resp.getWriter());
```

以上语句将结果数据和名称为 home 的模板交给 Thymeleaf 引擎进行处理。

注意:根据之前对 Thymeleaf 引擎的配置,home 模板将具体对应"/WEB-INF/templates/home.html"模板文件。

home.html 模板的文件内容如下:

```
<!DOCTYPE html>
<html xmlns:th="http://www.thymeleaf.org" lang="java">
<head>
    <title>Home</title>
    <meta http-equiv="Content-Type" content="text/html; charset=UTF-8" />
    <link rel="stylesheet" type="text/css"
        href="../../css/style.css" th:href="@{/css/style.css}"/>
</head>
<body>
    <h1>Hello world</h1>
    <p> from <label th:text="${name}"></label> ! </p>
</body>
</html>
```

注意其中的模板语句:

```
<link rel="stylesheet" type="text/css"
    href="../../css/style.css" th:href="@{/css/style.css}"/>
```

和

```
<p> from <label th:text="${name}"></label> ! </p>
```

Thymeleaf 引擎会扫描这个模板文档,结合传递到 Thymeleaf 上下文中的数据,Thymeleaf 会将这些模板标签进行数据置换。home.html 模板经过 Thymeleaf 引擎处理后,形成新的 HTML 文档内容如下:

```html
<!DOCTYPE html>
<html lang="java">
<head>
    <title>Home</title>
    <meta http-equiv="Content-Type" content="text/html; charset=UTF-8" />
    <link rel="stylesheet" type="text/css"
          href="/ch10_war_exploded/css/style.css"/>
</head>
<body>
    <h1>Hello world</h1>
    <p> from <label>Thymeleaf 使用举例</label> ! </p>
</body>
</html>
```

从这里可以看出，Thymeleaf 引擎正确处理模板和数据，产生了正确的最终 HTML 文档。运行这个程序，在浏览器地址栏输入 first，显示如图 10-3 所示的结果。

图 10-3　第一个 Thymeleaf 示例的运行结果

从 home.html 模板文档可以看出，Thymeleaf 其实就是将模板文档的模板表达式和模板标签替换为数据结果中的值。

Thymeleaf 工作原理观察

10.2　Thymeleaf 模板表达式

Thymeleaf 引擎支持多种方式的表达式，包括消息表达式♯{…}、变量表达式${…}、选择对象表达式＊{…}、URL 链接表达式@{…}等。通过这些表达式，可以将 Thymeleaf 上下文中的数据填充到 HTML 模板中。

10.2.1　消息表达式♯{…}

通过消息表达式可以将存储于外部资源文件中的文字填充到 HTML 模板文档中。外部资源文件必须与模板文件在同一个目录下，并且其文件名必须为"模板文件名_语言名.properties"。由于有"语言名"后缀，说明 Thymeleaf 是支持多语言的，当然，其中的"语言名"可以省略。如果资源文件中包含中文字符，需要在 IDEA 中修改文件的字符集。为此，在 IDEA 主界面选择 File→Settings 命令，在弹出的界面选择 Editor→File Encodings 选项，如图 10-4 所示。

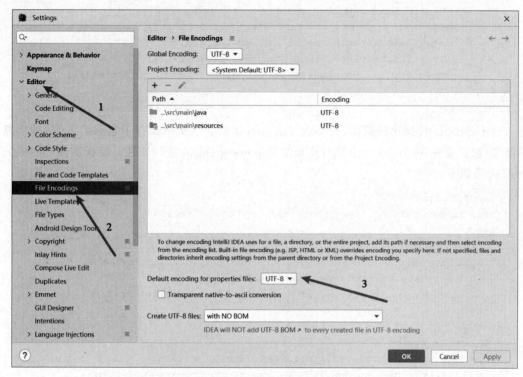

图 10-4　设置资源文件的字符集为 UTF-8

下面举例说明消息表达式的使用。这个例子中，名为 Message 的 Servlet 程序将创建 Thymeleaf 引擎，并委托引擎对名为 message.html 模板进行解析处理。Message.java 的代码如下。

```
package com.ttt.servlet;
import com.ttt.listener.ThymeleafEngine;
import jakarta.servlet.ServletException;
import jakarta.servlet.annotation.WebServlet;
import jakarta.servlet.http.HttpServlet;
import jakarta.servlet.http.HttpServletRequest;
import jakarta.servlet.http.HttpServletResponse;
import org.thymeleaf.TemplateEngine;
import org.thymeleaf.context.WebContext;
import org.thymeleaf.web.IWebExchange;
import org.thymeleaf.web.servlet.JakartaServletWebApplication;
import java.io.IOException;
@WebServlet("/message")
public class Message extends HttpServlet{
    @Override
    protected void doGet(HttpServletRequest req, HttpServletResponse resp)
      throws ServletException, IOException{
```

```
        TemplateEngine templateEngine = (TemplateEngine) getServletContext().
            getAttribute(ThymeleafEngine.THYMELEAF_ENGINE);
        IWebExchange webExchange = JakartaServletWebApplication
            .buildApplication(getServletContext())
            .buildExchange(req, resp);
        WebContext context = new WebContext(webExchange);
        context.setVariable("hello", "Thymeleaf使用举例");
        resp.setContentType("text/html; charset=UTF-8");
        templateEngine.process("message", context, resp.getWriter());
    }
}
```

这个Servlet代码比较简单,只是在Thymeleaf的上下文中保存了名称为hello的字符串信息,然后委托Thymeleaf引擎对模板文件message.html进行处理。模板文件message.html代码如下:

```
<!DOCTYPE html>
<html xmlns:th="http://www.thymeleaf.org" lang="java">
<head>
  <title>Home</title>
  <meta http-equiv="Content-Type" content="text/html; charset=UTF-8" />
  <link rel="stylesheet" type="text/css"
      href="../../css/style.css" th:href="@{/css/style.css}"/>
</head>
<body>
  <h1>Hello world</h1>
  <p> from <label th:text="${hello}"></label> ! </p>
  <p th:text="#{message.hello}">Welcome to our grocery store!</p>
  <p th:text="#{other}">Welcome to our grocery store!</p>
</body>
</html>
```

在模板文件中,使用语句:

```
<p th:text="#{message.hello}">Welcome to our grocery store!</p>
<p th:text="#{other}">Welcome to our grocery store!</p>
```

显示该模板对应的资源文件message.properties中的信息。message.properties的内容如下:

```
message.hello=你好!
other=A new message for example
```

运行这个程序,在浏览器地址栏输入"/message",显示如图10-5的界面。

10.2.2 变量表达式${...}

类似于JSP的EL表达式,使用Thymeleaf的变量表达式${...}可以访问Thymeleaf上下文中的信息。Thymeleaf提供了三个变量作用域。

(1) WebContext作用域:也就是Thymeleaf的页面作用域,只对当前页面有效,使用"${变量名}"或者"${#ctx.变量名}"可以访问WebContext作用域的变量。

图 10-5　消息表达式使用举例

语言名及其简称

（2）Session 作用域：也就是 HttpSession 作用域，对整个会话期有效，使用"${session.变量名}"可以访问 Session 作用域的变量。

（3）ServletContext 作用域：也就是 application 作用域，对整个应用程序有效，使用"${application.变量名}"可以访问 application 作用域额变量。

下面举例说明变量表达式${...}的使用。这个例子中，名为 Variables 的 Servlet 程序将创建 Thymeleaf 引擎，并委托引擎对名为 variables.html 模板进行解析处理，同时，为了说明变量的作用域，还编写了名为 Other 的 Servlet，这个 Servlet 只是将信息简单地委托给 variables.html 模板进行显示，以说明 Session 作用域和 Application 作用域的使用。Variables.java 代码如下。

```java
package com.ttt.servlet;
import com.ttt.listener.ThymeleafEngine;
import jakarta.servlet.ServletContext;
import jakarta.servlet.ServletException;
import jakarta.servlet.annotation.WebServlet;
import jakarta.servlet.http.HttpServlet;
import jakarta.servlet.http.HttpServletRequest;
import jakarta.servlet.http.HttpServletResponse;
import org.thymeleaf.TemplateEngine;
import org.thymeleaf.context.WebContext;
import org.thymeleaf.web.IWebExchange;
import org.thymeleaf.web.IWebSession;
import org.thymeleaf.web.servlet.JakartaServletWebApplication;
import java.io.IOException;
@WebServlet("/variables")
public class Variables extends HttpServlet{
    @Override
    protected void doGet(HttpServletRequest req, HttpServletResponse resp)
      throws ServletException, IOException{
        TemplateEngine templateEngine = (TemplateEngine) getServletContext()
          .getAttribute(ThymeleafEngine.THYMELEAF_ENGINE);
```

```java
            IWebExchange webExchange = JakartaServletWebApplication
                .buildApplication(getServletContext())
                .buildExchange(req, resp);
            WebContext context = new WebContext(webExchange);
            context.setVariable("hello", "来自 WebContext");
            IWebSession session = webExchange.getSession();
            session.setAttributeValue("hello", "来自 Session");
            ServletContext servletContext = req.getServletContext();
            servletContext.setAttribute("hello", "来自 ServletContext");
            resp.setContentType("text/html; charset=UTF-8");
            templateEngine.process("variables", context, resp.getWriter());
    }
}
```

这个 Servlet 只是简单地向 WebContext 对象、IWebSession 对象和 ServletContext 对象填写了一些属性值，然后将模板 variables.html 委托给 Thymeleaf 引擎进行处理。variables.html 的代码如下：

```html
<!DOCTYPE html>
<html xmlns:th="http://www.thymeleaf.org" lang="java">
<head>
  <title>Home</title>
  <meta http-equiv="Content-Type" content="text/html; charset=UTF-8" />
  <link rel="stylesheet" type="text/css"
        href="../../css/style.css" th:href="@{/css/style.css}"/>
</head>
<body>
  <h1>Hello world</h1>
  <p th:text="${hello}">place holder!</p>
  <p th:text="${#ctx.hello}">place holder!</p>
  <p th:text="${session.hello}">place holder!</p>
  <p th:text="${application.hello}">place holder!</p>
</body>
</html>
```

这个模板只是将保存在 WebContext 对象、IWebSession 对象和 ServletContext 对象中的信息显示在页面上。

为了观察 Thymeleaf 提供的三个变量作用域，再编写一个名为 Other 的 Servlet 程序，这个程序只是简单地获取 Thymeleaf 引擎，并委托它对模板 variables.html 进行处理。Other.java 的代码如下：

```java
package com.ttt.servlet;
import com.ttt.listener.ThymeleafEngine;
import jakarta.servlet.ServletException;
import jakarta.servlet.annotation.WebServlet;
import jakarta.servlet.http.HttpServlet;
import jakarta.servlet.http.HttpServletRequest;
import jakarta.servlet.http.HttpServletResponse;
import org.thymeleaf.TemplateEngine;
import org.thymeleaf.context.WebContext;
import org.thymeleaf.web.IWebExchange;
```

```
import org.thymeleaf.web.servlet.JakartaServletWebApplication;
import java.io.IOException;
@WebServlet("/other")
public class Other extends HttpServlet{
    @Override
    protected void doGet(HttpServletRequest req, HttpServletResponse resp)
      throws ServletException, IOException {
        TemplateEngine templateEngine = (TemplateEngine) getServletContext()
           .getAttribute(ThymeleafEngine.THYMELEAF_ENGINE);
        IWebExchange webExchange = JakartaServletWebApplication
           .buildApplication(getServletContext())
           .buildExchange(req, resp);
        WebContext context = new WebContext(webExchange);
        resp.setContentType("text/html; charset=UTF-8");
        templateEngine.process("variables", context, resp.getWriter());
    }
}
```

现在运行这个程序，在浏览器地址栏输入/variables，显示如图 10-6 所示的结果。

图 10-6　变量表达式使用举例

从图 10-6 可以看出，在未指明作用域时，使用变量表达式将默认从 WebContext 对象中获取属性值，因此，模板语句：

```
<p th:text="${hello}">place holder!</p>
<p th:text="${#ctx.hello}">place holder!</p>
```

将从 WebContext 对象中获取属性值，所以显示了一致的结果。而以下语句：

```
<p th:text="${session.hello}">place holder!</p>
<p th:text="${application.hello}">place holder!</p>
```

则从指定的作用域获取数据。当指定的作用域不存在指定的数据时则显示空串（注意：是空串""而不是 null）。现在在浏览器地址栏输入/other，显示如图 10-7 所示的界面。

图 10-7　变量表达式作用域使用举例

从结果可以看出，存储在 IWebSession 对象和 ServletContext 对象中的信息可以正常显示在页面上，而存储在 WebContext 对象的信息则不能访问。这说明存储在 IWebSession 对象和 ServletContext 对象中的信息是跨页面可以访问的，而存储在 WebContext 对象中的信息则不能跨页面访问。

10.2.3 选择对象表达式 *{...}

当要访问的变量是 POJO 对象的属性时，使用选择对象表达式 *{...} 可以简化代码编写。所谓选择对象，是指使用 Thymeleaf 的 <th:object> 标签指定一个对象，在后续对对象的属性的访问可以不用再指明对象名称。

下面举例说明选择对象表达式 *{...} 的使用。这个例子中，名为 Selection 的 Servlet 程序首先创建一个 Student 对象，然后创建 Thymeleaf 引擎，并委托引擎对名为 selection.html 模板进行解析处理。Selection.java 代码如下：

```java
package com.ttt.servlet;
import com.ttt.listener.ThymeleafEngine;
import com.ttt.pojo.Student;
import jakarta.servlet.ServletException;
import jakarta.servlet.annotation.WebServlet;
import jakarta.servlet.http.HttpServlet;
import jakarta.servlet.http.HttpServletRequest;
import jakarta.servlet.http.HttpServletResponse;
importorg.thymeleaf.TemplateEngine;
import org.thymeleaf.context.WebContext;
import org.thymeleaf.web.IWebExchange;
import org.thymeleaf.web.servlet.JakartaServletWebApplication;
import java.io.IOException;
@WebServlet("/selection")
public class Selection extends HttpServlet {
    @Override
    protected void doGet(HttpServletRequest req, HttpServletResponse resp)
      throws ServletException, IOException{
        TemplateEngine templateEngine = (TemplateEngine) getServletContext()
            .getAttribute(ThymeleafEngine.THYMELEAF_ENGINE);
        IWebExchange webExchange = JakartaServletWebApplication
            .buildApplication(getServletContext())
            .buildExchange(req, resp);
        WebContext context = new WebContext(webExchange);
        Student student = new Student("张三", 20, "13800138000",
            "一个学习优秀道德优秀的学生");
        context.setVariable("student", student);
        resp.setContentType("text/html; charset=UTF-8");
        templateEngine.process("selection", context, resp.getWriter());
    }
}
```

这个 Servlet 程序创建了 Student 对象并存储到 WebContext 对象中,然后委托 Thymeleaf 引擎进行处理。POJO 类 Student.java 的代码如下:

```java
package com.ttt.pojo;
public class Student{
    private final String name;
    private final int age;
    private final String phone;
    private final String memo;
    public Student(String name, int age, String phone, String memo){
        this.name = name;
        this.age = age;
        this.phone = phone;
        this.memo = memo;
    }
    public String getName(){
        return name;
    }
    public int getAge(){
        return age;
    }
    public String getPhone(){
        return phone;
    }
    public String getMemo(){
        return memo;
    }
    @Override
    public String toString(){
        return "Student{" +
            "name='" + name + '\'' +
            ", age=" + age +
            ", phone='" + phone + '\'' +
            ", memo='" + memo + '\'' +
            '}';
    }
}
```

selection.html 模板代码如下:

```html
<!DOCTYPE html>
<html xmlns:th="http://www.thymeleaf.org" lang="java">
<head>
  <title>Home</title>
  <meta http-equiv="Content-Type" content="text/html; charset=UTF-8" />
  <link rel="stylesheet" type="text/css"
        href="../../css/style.css" th:href="@{/css/style.css}"/>
</head>
<body>
  <h1>Hello world</h1>
  <div th:object="${student}">
```

```html
  <p>姓名：<span th:text="*{name}">a</span>.</p>
  <p>年龄：<span th:text="*{age}">b</span>.</p>
  <p>联系电话：<span th:text="*{phone}">c</span>.</p>
  <p>简介：<span th:text="*{memo}">c</span>.</p>
</div>
<h1>等价于</h1>
<p>姓名：<span th:text="${student.name}">a</span>.</p>
<p>年龄：<span th:text="${student.age}">b</span>.</p>
<p>联系电话：<span th:text="${student.phone}">c</span>.</p>
<p>简介：<span th:text="${student.memo}">c</span>.</p>
</body>
</html>
```

在 selection.html 模板代码中使用以下语句：

```html
<div th:object="${student}">
  <p>姓名：<span th:text="*{name}">a</span>.</p>
  <p>年龄：<span th:text="*{age}">b</span>.</p>
  <p>联系电话：<span th:text="*{phone}">c</span>.</p>
  <p>简介：<span th:text="*{memo}">c</span>.</p>
</div>
```

以上语句通过选择对象表达式后再访问对象的属性值，这段代码与如下代码是完全等价的：

```html
<p>姓名：<span th:text="${student.name}">a</span>.</p>
<p>年龄：<span th:text="${student.age}">b</span>.</p>
<p>联系电话：<span th:text="${student.phone}">c</span>.</p>
<p>简介：<span th:text="${student.memo}">c</span>.</p>
```

也就是说，选择对象表达式完全可以被变量表达式取代。运行这个程序，在浏览器地址栏输入"/selection"，显示如图 10-8 所示的结果。

图 10-8　选择对象表达式应用示例

10.2.4　URL 链接表达式 @{...}

URL 链接表达式 @{...} 用于构建完整的 URL 链接地址，用以取代 HTML 同名属性中的 URL 链接地址。例如，对于 HTML 的<a>标签的 href 属性，可以使用 Thymeleaf 的 th:href 属性指定最终有效的 URL 地址；对于 HTML 的<img>标签的 src 属性，可以使用 Thymeleaf 的 th:src 属性指定最终有效的 URL 地址等。不仅如此，URL 链接表达式 @{...}还可以附加链接参数，并可以使用 Thymeleaf 的其他表达式模板产生动态 URL 地址。

下面举例说明 URL 链接表达式@{...}的使用。这个例子中，名为 LinkURL 的 Servlet 程序首先设置一些基本的属性数据，然后创建 Thymeleaf 引擎，并委托引擎对名为 linkurl.html 模板进行解析处理。LinkURL.java 的代码如下：

```java
package com.ttt.servlet;
import com.ttt.listener.ThymeleafEngine;
import jakarta.servlet.ServletException;
import jakarta.servlet.annotation.WebServlet;
import jakarta.servlet.http.HttpServlet;
import jakarta.servlet.http.HttpServletRequest;
import jakarta.servlet.http.HttpServletResponse;
import org.thymeleaf.TemplateEngine;
import org.thymeleaf.context.WebContext;
import org.thymeleaf.web.IWebExchange;
import org.thymeleaf.web.servlet.JakartaServletWebApplication;
import java.io.IOException;
@WebServlet("/linkurl")
public class LinkURL extends HttpServlet{
    @Override
    protected void doGet(HttpServletRequest req, HttpServletResponse resp)
        throws ServletException, IOException{
        req.setCharacterEncoding("UTF-8");
        String height = req.getParameter("height");
        String width = req.getParameter("width");
        TemplateEngine templateEngine = (TemplateEngine) getServletContext()
            .getAttribute(ThymeleafEngine.THYMELEAF_ENGINE);
        IWebExchange webExchange = JakartaServletWebApplication
            .buildApplication(getServletContext())
            .buildExchange(req, resp);
        WebContext context = new WebContext(webExchange);
        context.setVariable("image", "/images/wheat.jpg");
        context.setVariable("height", height);
        context.setVariable("width", width);
        resp.setContentType("text/html; charset=UTF-8");
        templateEngine.process("linkurl", context, resp.getWriter());
    }
}
```

该程序首先获取名为 height 和 width 的请求参数，然后获取 Thymeleaf 引擎对象并委托 linkurl.html 模板生成客户端文档。linkurl.html 模板代码如下：

```html
<!DOCTYPE html>
<html xmlns:th="http://www.thymeleaf.org" lang="java">
```

```html
<head>
  <title>Home</title>
  <meta http-equiv="Content-Type" content="text/html; charset=UTF-8" />
  <link rel="stylesheet" type="text/css"
        href="../../css/style.css" th:href="@{/css/style.css}"/>
</head>
<body>
  <h1>Hello world</h1>
  <img src="" th:src="@{/images/wheat.jpg}" th:height="${height}"
              th:width="${width}" alt=""/>
  <img src="" th:src="@{${image}}" alt="">
  <p>
    <a href="" th:href="@{/linkurl(height=200,width=200)}">单击访问另一个图片</a>
  </p>
</body>
</html>
```

留意代码中加黑显示的语句。在这些语句中，都使用了 Thymeleaf 的 URL 链接表达式以构建完整的 URL 地址。并且以下语句：

```html
<img src="" th:src="@{${image}}" alt="">
```

使用了 Thymeleaf 的变量表达式用以构造动态 URL。以下语句：

```html
<a href="" th:href="@{/linkurl(height=200,width=200)}">单击访问另一个图片</a>
```

用以构造带参数的 URL 请求，这条语句经过 Thymeleaf 引擎处理后将产生如下语句：

```
http://localhost:8080/ch10_war_exploded/linkurl?height=200&width=200
```

这是期望的结果。运行这个程序，在浏览器地址栏输入"/linkurl"，显示如图 10-9 所示的界面。

图 10-9　URL 链接表达式应用示例

在图 10-9 的界面中单击"单击访问另一个图片",将显示如图 10-10 所示的界面。

图 10-10　另一个 URL 链接表达式应用示例界面

10.3　Thymeleaf 的字面常量和运算符

类似于其他标记/脚本语言,Thymeleaf 支持字面常量、算术运算、关系运算和逻辑运算、字符串操作和条件运算等。

10.3.1　字面常量

Thymeleaf 支持的字面常量包括字符串字面量、数值字面量、逻辑字面量和 null 字面量。其中,数值字面量包括整数字面量、浮点字面量,例如,整数 1234 和浮点 2345.12 等;字符串字面量是使用单引号引起来的字符值,例如,'Hello,world'、'你最近在忙什么?'等;逻辑字面量包括 true 和 false,分别表示逻辑"真"和逻辑"假";null 字面量就是空对象。

10.3.2　字符串操作

Thymeleaf 支持字符串连接操作和字符串替换操作,其中,使用符号"＋"完成字符串连接操作,使用"| |"完成字符串替换操作。例如,以下语句:

`<span th:text="'The name of the user is ' + ${user.name}">`

将字符串'The name of the user is'与变量表达式${user.name}的值做连接操作。以下语句:

`<span th:text="|Welcome to our application, ${user.name}|">`

则将字符串'Welcome to our application,${user.name}'中的变量表达式${user.name}替换为其实际值。本质上,它等价于上一句的字符串连接操作。

10.3.3 算术运算、关系运算和逻辑运算

Thymeleaf 支持常用的算术运算符包括＋、－、＊、/、％等，支持的关系运算包括＞、＞＝、＜、＜＝、＝＝、!＝等，用于判断两个值的关系，支持的逻辑运算包括 and、or、not。

10.3.4 条件运算符

Thymeleaf 的条件运算符类似于其他编程语言的"?:"运算符。其一般形式是"条件？值 1：值 2"，也就是说，当"条件"为 true 时，取"值 1"的值，否则取"值 2"的值。在 Thymeleaf 中，条件运算符中的"值 2"是可以省略的，在这种情况下，当"条件"为 false 时，取 null 值。

10.3.5 字面常量和运算符使用举例

下面举例说明字面常量和运算符的使用。这个例子中，名为 Literal 的 Servlet 程序首先设置一些基本的属性数据，然后创建 Thymeleaf 引擎，并委托引擎对名为 literal.html 模板进行解析处理。Literal.java 的代码如下：

```java
package com.ttt.servlet;
import com.ttt.listener.ThymeleafEngine;
import jakarta.servlet.ServletException;
import jakarta.servlet.annotation.WebServlet;
import jakarta.servlet.http.HttpServlet;
import jakarta.servlet.http.HttpServletRequest;
import jakarta.servlet.http.HttpServletResponse;
import org.thymeleaf.TemplateEngine;
import org.thymeleaf.context.WebContext;
import org.thymeleaf.web.IWebExchange;
import org.thymeleaf.web.servlet.JakartaServletWebApplication;
import java.io.IOException;
@WebServlet("/literal")
public class Literal extends HttpServlet{
    @Override
    protected void doGet(HttpServletRequest req, HttpServletResponse resp)
      throws ServletException, IOException {
        req.setCharacterEncoding("UTF-8");
        TemplateEngine templateEngine = (TemplateEngine) getServletContext()
            .getAttribute(ThymeleafEngine.THYMELEAF_ENGINE);
        IWebExchange webExchange = JakartaServletWebApplication
            .buildApplication(getServletContext())
            .buildExchange(req, resp);
        WebContext context = new WebContext(webExchange);
        context.setVariable("a", 100);
        context.setVariable("b", 200);
        context.setVariable("name", "张三");
```

```
        context.setVariable("flag", true);
        resp.setContentType("text/html; charset=UTF-8");
        templateEngine.process("literal", context, resp.getWriter());
    }
}
```

在这段代码中,首先在 WebContetx 中设置了一些属性,然后得到 Thymeleaf 引擎对象并委托对模板 literal.html 进行处理。literal.html 的代码如下。

```
<!DOCTYPE html>
<html xmlns:th="http://www.thymeleaf.org" lang="java">
<head>
  <title>Home</title>
  <meta http-equiv="Content-Type" content="text/html; charset=UTF-8" />
  <link rel="stylesheet" type="text/css"
        href="../../css/style.css" th:href="@{/css/style.css}"/>
</head>
<body>
  <h1>Hello world</h1>
  <div th:text="${a} + ${b}"></div>
  <div th:text="${a} % ${b}"></div>
  <div th:text="${a} >= ${b}"></div>
  <br>
<div th:text="|我的名字叫${name}|"></div>
<div th:text="'我的名字叫' + ${name}"></div>
  <br>
  <div th:text="${a != null} and ${b != null}"></div>
  <div th:text="${a == 100} and ${b != 200}"></div>
  <br>
  <div th:text="${name=='张三'}?'张三':'不是张三'"></div>
  <div th:if="${name=='张三'}">张三</div>
  <br>
  <div th:with="isEven=(${a} % 2  == 0)">
    <p th:text="${isEven}"></p>
  </div>
</body>
</html>
```

运行这个程序,在浏览器地址栏输入"/linteral",将显示如图 10-11 所示的结果。

图 10-11　字面常量和运算符使用举例

从运行结果可以看出，Thymeleaf 进行了正确的运算和结果显示。在这个例子中，使用了 Thymeleaf 的一些常用的属性，如 th:text、th:if、th:with 等，下面对 Thymeleaf 常用的属性进行介绍。

10.4 Thymeleaf 常用属性及其使用

Thymeleaf 通过名称空间 xmlns:th="http://www.thymeleaf.org" 为 HTML 扩展了很多属性，其中最为常用的属性包括 th:text、th:utext、th:attr、th:with、th:if、th:switch、th:each 等。其中，th:if、th:switch、th:each 涉及条件控制和迭代，这些内容将在 10.5 节介绍。本节对 Thymeleaf 的常用属性进行介绍。

10.4.1 使用 th:text、th:utext 和内联属性输出文字

th:text 和 th:utext 属性对其属性值进行处理，并取代其属主 HTML 标签的内容。th:text 和 th:utext 都用于展示纯文本，但是，th:text 会对特殊字符进行转义，而 th:utext 则不会对特殊字符做转义处理。使用两个中括号括起来的 Thymeleaf 表达式称为内联。例如，如果名称为 hello 的变量的值为"你好,世界"，那么，如下的模板语句：

```
<p th:text="${hello}">123</p>
<p th:utext="${hello}">123</p>
<p>在这里使用内联：[[${hello}]]</p>
```

经过 Thymeleaf 引擎处理后，生成如下最终的 HTML 语句：

```
<p>&lt;b&gt;你好,世界 &lt;/b&gt;</p>
<p><b>你好,世界</b></p>
<p>在这里使用内联：&lt;b&gt;你好,世界 &lt;/b&gt;</p>
```

另外，Thymeleaf 的内联可以在 JavaScript 中访问属性变量的值。例如，如下的代码在 JavaScript 中访问内联的 Thymeleaf 变量的值：

```
<script type="text/javascript" th:inline="javascript">
    var max = [[${hello}]];
    console.log(max);
</script>
```

这段代码将在浏览器的控制台输出"你好,世界"。

10.4.2 使用 th:with 属性定义局部变量

Thymeleaf 的 th:with 属性用于定义局部变量。例如，如果名称为 a 的变量的值为 100，如下的模板语句：

```
<div th:with="isEven=(${a} % 2 == 0)">
  <p th:text="${isEven}"></p>
</div>
```

定义了名为 isEven 的局部变量，其取值为属性变量 a 除以 2 的余数是否为 0 这个逻辑值。由于 a 的值是 100，所以 isEven 的值为 true，因此，这个模板将显示 true 到浏览器上。

10.4.3 使用 th:attr 属性设置 HTML 标签的属性值

Thymeleaf 的 th:attr 属性用于修改或设置 HTML 标签属性的值。th:attr 属性使用额一般格式如下：

<html 标签名 th:attr="标签属性名=标签属性值" html 其他属性…>

例如，如果 subscribe 属性变量的值为"/Subscribe"，外部资源 subscribe.submit 的值为"提交"，则模板代码：

```
<form action="subscribe.html" th:attr="action=@{/subscribe}">
  <fieldset>
    <input type="text" name="email" />
    <input type="submit" value="Subscribe!" th:attr="value=#{subscribe.submit}"/>
  </fieldset>
</form>
```

被 Thymeleaf 引擎处理后，最终的 HTML 代码如下：

```
<form action="应用程序上下文/Subscribe">
  <fieldset>
    <input type="text" name="email" />
    <input type="submit" value="提交"/>
  </fieldset>
</form>
```

也可以使用 th:attr 一次性设置或修改多个 HTML 标签的属性值。例如，如下的模板代码：

```
<img src="../../images/gtvglogo.png"
  th:attr="src=@{/images/gtvglogo.png},title=#{logo},alt=#{logo}" />
```

将一次性设置/修改 img 标签的 src 属性、title 属性和 alt 属性的值。

在可以使用 th:attr 设置/修改 HTML 属性值的同时，Thymeleaf 为了方便编写模板代码，为常用的 HTML 标签属性定义了专门的 Thymeleaf 属性。例如，对于模板代码：

```
<img src="../../images/gtvglogo.png"
  th:attr="src=@{/images/gtvglogo.png},title=#{logo},alt=#{logo}" />
```

完全可以使用如下等价的模板代码表示：

```
<img src="../../images/gtvglogo.png"
  th:src="@{/images/gtvglogo.png}" title="#{logo}" alt="#{logo}" />
```

Thymeleaf 已经为常用的 HTML 标签属性定义了简约属性，这些简约属性包括 th:abbr、th:accept、th:accept-charset、th:accesskey、th:action、th:align、th:alt、th:archive、th:audio、th:autocomplete、th:axis、th:background、th:bgcolor、th:border 等。关于完整的 Thymeleaf 简约属性，可参见 Thymeleaf 官网。对于这些属性的使用，在编写模板代码时，

可以通过 IDEA 的提示进行选择即可，如图 10-12 所示。

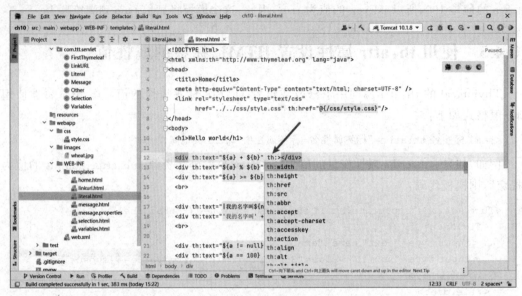

图 10-12　使用 IDEA 的提示选择 Thymeleaf 属性

10.5　Thymeleaf 的条件控制和迭代

Thymeleaf 提供了条件控制属性 th:if、th:switch/th:case，也提供了数据迭代属性 th:each 对模板数据进行控制。

10.5.1　th:each 迭代的使用

Thymeleaf 引擎可以在模板中对如下类型的变量进行迭代，这些可迭代的类型包括数组、java.util.List、java.util.Map、java.util.Iterator、java.util.Enumeration、java.util.Iterable、java.util.stream.Stream。通过 th:each 迭代，可以对集合数据变量的每个元素进行处理。

下面举例说明 th:each 迭代的使用。这个例子中，名为 Foreach 的 Servlet 程序首先创建一个包含 Student 对象的 List，然后创建 Thymeleaf 引擎，并委托引擎对名为 foreach.html 模板进行解析处理。Foreach.java 代码如下：

```
package com.ttt.servlet;
import com.ttt.listener.ThymeleafEngine;
import com.ttt.pojo.Student;
import jakarta.servlet.ServletException;
import jakarta.servlet.annotation.WebServlet;
import jakarta.servlet.http.HttpServlet;
import jakarta.servlet.http.HttpServletRequest;
import jakarta.servlet.http.HttpServletResponse;
import org.thymeleaf.TemplateEngine;
import org.thymeleaf.context.WebContext;
```

```java
import org.thymeleaf.web.IWebExchange;
import org.thymeleaf.web.servlet.JakartaServletWebApplication;
import java.io.IOException;
import java.util.LinkedList;
import java.util.List;
@WebServlet("/foreach")
public class Foreach extends HttpServlet{
    @Override
    protected void doGet(HttpServletRequest req, HttpServletResponse resp)
      throws ServletException, IOException{
        TemplateEngine templateEngine = (TemplateEngine) getServletContext()
            .getAttribute(ThymeleafEngine.THYMELEAF_ENGINE);
        IWebExchange webExchange = JakartaServletWebApplication
            .buildApplication(getServletContext())
            .buildExchange(req, resp);
        WebContext context = new WebContext(webExchange);
        List<Student> list = new LinkedList<>();
        Student stu1 = new Student("张三", 20, "13555555555", "好同学张三");
        Student stu2 = new Student("李四", 21, "13566666666", "好同学李四");
        Student stu3 = new Student("王五", 22, "135777777777", "好同学王五");
        Student stu4 = new Student("马六", 23, "13588888888", "好同学马六");
        list.add(stu1);list.add(stu2);list.add(stu3);list.add(stu4);
        context.setVariable("students", list);
        resp.setContentType("text/html; charset=UTF-8");
        templateEngine.process("foreach", context, resp.getWriter());
    }
}
```

这个程序创建包含四个 Student 对象的 List，并设置到 WebContext 对象中，然后委托 Thymeleaf 引擎对 foreach.html 模板进行处理。foreach.html 的代码如下：

```html
<!DOCTYPE html>
<html xmlns:th="http://www.thymeleaf.org" lang="java">
<head>
  <title>Home</title>
  <meta http-equiv="Content-Type" content="text/html; charset=UTF-8" />
  <link rel="stylesheet" type="text/css"
        href="../../css/style.css" th:href="@{/css/style.css}"/>
</head>
<body>
  <h1>学生列表</h1>
  <table>
    <tr>
      <th>姓名</th>
      <th>年龄</th>
      <th>联系电话</th>
      <th>简介</th>
    </tr>
    <tr th:each="stu : ${students}">
      <td th:text="${stu.name}">A</td>
      <td th:text="${stu.age}">B</td>
      <td th:text="${stu.phone}">C</td>
      <td th:text="${stu.memo}">C</td>
    </tr>
```

```
    </table>
  </body>
</html>
```

在这个模板代码中,通过以下语句:

```
<tr th:each="stu : ${students}">
    <td th:text="${stu.name}">A</td>
    <td th:text="${stu.age}">B</td>
    <td th:text="${stu.phone}">C</td>
    <td th:text="${stu.memo}">C</td>
</tr>
```

对列表 students 对象的每个元素进行迭代,针对每个元素产生 HTML 的 tr 标签,并显示每个 Student 对象的属性值。其中,以下语句:

```
<tr th:each="stu : ${students}">
```

获取 students 对象的每个元素,并赋值到局部变量 stu 中,进而可在 th:foreach 中通过这个局部变量访问每个 Student 对象的值。运行这个程序,在浏览器地址栏输入/foreach,显示如图 10-13 所示的结果。

图 10-13 th:foreach 迭代使用举例

为了便于对迭代的过程进行管理,Thymeleaf 对 th:each 迭代过程提供了状态变量,这个状态变量包括如下属性。

- index:当前迭代元素的索引号,从 0 开始。
- count:当前迭代元素的索引号,从 1 开始。
- size:被迭代的集合变量的元素总数。
- current:被当前被迭代元素的变量名。
- even/odd:当前被迭代的元素是否处于偶数/奇数位置。
- first:当前被迭代的元素是否是第一个元素。
- last:当前被迭代的元素是否是最后一个元素。

为了观察状态变量的使用,现在修改 foreach.html 模板代码如下:

```
<!DOCTYPE html>
<html xmlns:th="http://www.thymeleaf.org" lang="java">
<head>
    <title>Home</title>
    <meta http-equiv="Content-Type" content="text/html; charset=UTF-8" />
    <link rel="stylesheet" type="text/css"
          href="../../css/style.css" th:href="@{/css/style.css}"/>
</head>
```

```html
<body>
  <h1>学生列表</h1>
  <table>
    <tr>
      <th>姓名</th>
      <th>年龄</th>
      <th>联系电话</th>
      <th>简介</th>
    </tr>
    <tr th:each="stu,stat : ${students}" th:class="${stat.odd}? 'odd'">
      <td th:text="${stu.name}">A</td>
      <td th:text="${stu.age}">B</td>
      <td th:text="${stu.phone}">C</td>
      <td th:text="${stu.memo}">C</td>
    </tr>
  </table>
</body>
</html>
```

在这个新的模板代码中使用如下语句：

```
<tr th:each="stu,stat : ${students}" th:class="${stat.odd}? 'odd'">
```

命名状态变量为 stat，进而可以通过 stat 变量为不同的 Student 对象采用不同的背景表示。运行这个程序，将显示如图 10-14 所示的结果。

图 10-14　使用 th:each 的状态变量

Thymeleaf 的 th:each 迭代观察

10.5.2　th:if 和 th:unless 条件控制的使用

th:if 条件控制用于控制其属主 HTML 标签是否输出到 Thymeleaf 引擎的结果文档中：当 th:if 的结果为 true 时输出，否则，其属主 HTML 标签不会出现在结果文档中。例如，如果名称为 flag 的属性变量的值为 true，则下面的模板语句：

```
<p th:if=${flag}>你好,世界</p>
```

将在页面上显示"你好,世界"。如果 flag 的 false,则不会显示这段文字。

th:unless 条件控制也用于控制其属主 HTML 标签是否输出到 Thymeleaf 引擎的结果文档中:当 th:unless 的结果值为 false 时输出,否则,当结果只为 true 时,其属主 HTML 标签不会出现在结果文档中。举例来说,以下语句:

```
<p th:if=${flag}>你好,世界</p>
```

等价于:

```
<p th:unless=${not flag}>你好,世界</p>
```

10.5.3　th:switch/th:case 多分支控制的使用

th:switch/th:case 用于多分支控制,作用类似于其他编程语言中的 switch...case...default 语句。例如:

```
<div th:switch="${user.role}">
    <p th:case="'admin'">管理员</p>
    <p th:case="manager">经理</p>
    <p th:case="*">普通用户</p>
</div>
```

这段模板语句检查变量 user 的 role 属性,如果其值为 admin,则输出"管理员";如果其值为 manager,则输出"经理";否则输出"普通用户"。

注意:th:case="*"相当于 default 语句。

JSP vs Thymeleaf

10.6　Thymeleaf 工具类及其使用

为了便于在 HTML 模板中对变量进行访问和操作,Thymeleaf 提供了一系列工具类对象,这些对象及其作用如表 10-1 所示。

表 10-1　Thymeleaf 的工具类及其作用

序号	工具类名称	作　　用
1	#execInfo	获取被处理的模板及其相关信息
2	#messages	提供了一系列用于访问外部消息资源的方法,类似于 Thymeleaf 表达式的#{…}的作用
3	#uris	提供了转义 URL/URI 的相关方法

续表

序号	工具类名称	作用
4	#conversions	提供了配置转换方法
5	#dates	提供了类似于 java.util.Date 的方法,用于对日期进行处理
6	#calendars	#dates 工具对象的别名
7	#numbers	提供了对数字进行格式化的相关方法
8	#strings	提供了对字符串进行操作的方法
9	#bools	提供了多 boolean 类型进行操作的方法
10	#arrays	提供了一系列对数组进行操作的方法
11	#lists	提供了对列表进行操作的方法
12	#sets	提供了对 set 集合进行操作的方法
13	#maps	提供了对 map 进行操作的方法
14	#aggregates	提供了针对数组或者集合类型进行统计的相关方法
15	#ids	用于在模板中产生 HTML 标签的 id 属性

下面举例说明 Thymeleaf 工具类的使用。这个例子中,名为 ThymeUtil 的 Servlet 程序首先创建一个包含 Student 对象的 List、一个字符串数组和一些普通变量,然后创建 Thymeleaf 引擎,并委托引擎对名为 thymeutil.html 模板进行解析处理。ThymeUtil.java 代码如下:

```java
package com.ttt.servlet;
import com.ttt.listener.ThymeleafEngine;
import com.ttt.pojo.Student;
import jakarta.servlet.ServletException;
import jakarta.servlet.annotation.WebServlet;
import jakarta.servlet.http.HttpServlet;
import jakarta.servlet.http.HttpServletRequest;
import jakarta.servlet.http.HttpServletResponse;
import org.thymeleaf.TemplateEngine;
import org.thymeleaf.context.WebContext;
import org.thymeleaf.web.IWebExchange;
import org.thymeleaf.web.servlet.JakartaServletWebApplication;
import java.io.IOException;
import java.util.*;
@WebServlet("/thymeutil")
public class ThymeUtil extends HttpServlet {
    @Override
    protected void doGet(HttpServletRequest req, HttpServletResponse resp)
      throws ServletException, IOException{
        TemplateEngine templateEngine = (TemplateEngine) getServletContext()
            .getAttribute(ThymeleafEngine.THYMELEAF_ENGINE);
        IWebExchange webExchange = JakartaServletWebApplication
            .buildApplication(getServletContext())
            .buildExchange(req, resp);
        WebContext context = new WebContext(webExchange);
        context.setVariable("url", "http://www.example.com/index.html?p1=中国
```

```java
            &p2=GZ");
            context.setVariable("date", new Date());
            context.setVariable("salary", 2234.5678f);
            context.setVariable("name", "张三,别名张老三");
            context.setVariable("flag", true);
            List<Student> list = new LinkedList<>();
            Student stu1 = new Student("张三", 20, "13555555555", "好同学张三");
            Student stu2 = new Student("李四", 21, "13566666666", "好同学李四");
            Student stu3 = new Student("王五", 22, "13577777777", "好同学王五");
            Student stu4 = new Student("马六", 23, "13588888888", "好同学马六");
            list.add(stu1);
            list.add(stu2);
            list.add(stu3);
            list.add(stu4);
            context.setVariable("students", list);
            Map<String, Student> map = new LinkedHashMap<>();
            map.put("1", stu1);
            map.put("2", stu2);
            map.put("3", stu3);
            context.setVariable("map", map);
            resp.setContentType("text/html; charset=UTF-8");
            templateEngine.process("thymeutil", context, resp.getWriter());
    }
}
```

thymeutil.html 模板的代码如下：

```html
<!DOCTYPE html>
<html xmlns:th="http://www.thymeleaf.org" lang="java">
<head>
  <title>Home</title>
  <meta http-equiv="Content-Type" content="text/html; charset=UTF-8" />
  <link rel="stylesheet" type="text/css"
        href="../../css/style.css" th:href="@{/css/style.css}"/>
</head>
<body>
<h1>列表</h1>
  <p th:text="${#execInfo.processedTemplateName}"></p>
  <p th:text="${#uris.escapeQueryParam(url)}"></p>
  <p th:text="${#dates.format(date, 'yyyy-MM-dd HH:mm')}"></p>
  <p th:text="${#numbers.formatDecimal(salary, 10, 2)}"></p>
  <p th:text="${#strings.isEmpty(name)}"></p>
  <p th:text="${#strings.substring(name, 3, 5)}"></p>
  <p th:text="${#bools.isTrue(flag)}"></p>
  <p th:text="${#lists.size(students)}"></p>
  <p th:text="${#maps.isEmpty(map)}"></p>
</body>
</html>
```

现在运行这个程序，在浏览器地址栏输入"/thymeutil"，显示如图 10-15 所示的结果。

从图 10-15 的运行结果可以看出，Thymeleaf 的工具类正确地解析了变量的值。在

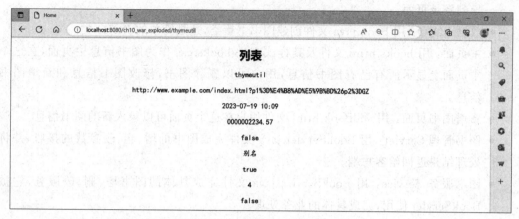

图 10-15　Thymeleaf 工具类的使用

IDEA 开发环境中使用 Thymeleaf 工具类时，IDEA 会给出工具类的方法和具体作用，可以参考提示使用 Thymeleaf 的工具类处理变量数据。

10.7　案例：图书信息管理系统

作为对 Servlet 技术和 Thymeleaf 展示技术的综合使用，本案例程序设计一个图书信息管理系统，用于对图书信息进行管理，具体要求同 9.6 的案例要求。但是，这里要求使用 Thymeleaf 作为页面表示技术。

10.7.1　案例目标

编写一个简单的图书管理系统，图书信息保存在数据库中。每本图书包含如下基本信息：名称、出版社、作者、书籍简介、书籍封面图片。在主页中，显示已经录入的所有图书信息，对已有的图书可以进行编辑，包括删除图书及修改图书信息，同时，在主页面中还包括一个"新增"按钮，单击该按钮，可以新增图书到数据库中。

10.7.2　案例分析

在可以对图书信息进行管理之前，需要首先登录到系统中，并且对于未登录的访问，系统将自动将访问引导到登录页面。因此，需要设计和编写如下程序文件。

- 入口：用 Entry.java 文件直接导向 login.html。
- 登录页面：用 login.html 文件及其样式文件 login.css。
- 登录服务 Servlet：用 Login.java 文件完成登录验证并创建 Session。登录成功后直接跳转到主页面，登录失败则再次显示登录页面。
- 登录检查过滤器：用 MyFilter.java 文件放行登录用户的访问，对未登录用户直接引导到登录页面。
- 数据 POJO：用 Book.java 文件创建图书书籍类，这是简单的 POJO 类。
- 主页面：用 books.html 文件及其样式文件 books.css 作为图书信息主页面，在这个主页面上显示所有已有图书信息，用户可以删除图书，修改图书信息和新增图书信息。
- 新增图书页面：用 addBook.html 文件用户在这个页面可以录入新的图书信息。
- 图书管理 Servlet：用 BookServlet.java 文件完成图书的增、删、改参数的接收，并将处理结果返回给客户端。
- 图书服务 Service：用 BookService.java 文件完成具体的图书增、删、改服务，它被 BookServlet 使用，完成具体的业务处理。
- 图书数据 DAO：用 BookDAO.java 文件完成具体数据信息的数据库操作。
- 数据库连接池：用 DruidUtil.java 文件及其配置文件 druid.properties 完成数据库连接池的管理。

10.7.3 案例实施

首先创建数据库及相关数据表。在 MySQL 下创建 dbbooks 数据库和 book 数据表，相关语句如下：

```sql
CREATE SCHEMA 'dbbooks' DEFAULT CHARACTER SET utf8mb4;
CREATE TABLE 'dbbooks'.'book'(
  'id' INT NOT NULL AUTO_INCREMENT,
  'name' VARCHAR(45) NOT NULL,
  'publisher' VARCHAR(200) NOT NULL,
  'author' VARCHAR(45) NOT NULL,
  'memo' TEXT(2000) NULL,
  'cover' LONGBLOB NULL,
  PRIMARY KEY ('id'));
```

（1）Entry.java 文件的代码。

Entry.java 文件的代码如下：

```java
package com.ttt.servlet;
import com.ttt.listener.ThymeleafEngine;
import jakarta.servlet.ServletException;
import jakarta.servlet.annotation.WebServlet;
import jakarta.servlet.http.HttpServlet;
import jakarta.servlet.http.HttpServletRequest;
import jakarta.servlet.http.HttpServletResponse;
import org.thymeleaf.TemplateEngine;
import org.thymeleaf.context.WebContext;
import org.thymeleaf.web.IWebExchange;
import org.thymeleaf.web.servlet.JakartaServletWebApplication;
import java.io.IOException;
@WebServlet("/entry")
public class Entry extends HttpServlet{
```

```java
    @Override
    protected void doGet(HttpServletRequest req, HttpServletResponse resp)
        throws ServletException, IOException{
        TemplateEngine templateEngine = (TemplateEngine) getServletContext()
            .getAttribute(ThymeleafEngine.THYMELEAF_ENGINE);
        IWebExchange webExchange = JakartaServletWebApplication
            .buildApplication(getServletContext())
            .buildExchange(req, resp);
        WebContext context = new WebContext(webExchange);
        context.setVariable("message", "");
        resp.setContentType("text/html; charset=UTF-8");
        templateEngine.process("login", context, resp.getWriter());
    }
}
```

(2) login.html 模板代码和 login.css 样式代码。

login.html 模板代码如下：

```html
<!DOCTYPE html>
<html xmlns:th="http://www.thymeleaf.org" lang="java">
<head>
  <title>图书信息管理</title>
  <meta http-equiv="Content-Type" content="text/html; charset=UTF-8" />
  <link rel="stylesheet" type="text/css" th:href="@{/css/login.css}"/>
</head>
<body>
<div>
  <h3>[[${message}]]</h3>
  <form enctype="application/x-www-form-urlencoded" method="POST"
      target="_self" action="./Login" th:action="@{/Login}">
    <label for="username">用户名</label>
    <input type="text" id="username" name="username" placeholder="用户名...">
    <label for="password">密码</label>
    <input type="password" id="password" name="password" placeholder="密
        码...">
    <input type="submit" value="提交">
  </form>
</div>
</body>
</html>
```

login.css 样式代码如下：

```css
div{
    width: 600px;
    margin: 100px auto;
}
input[type=text], input[type=password]{
    width: 100%;
    padding: 12px 20px;
    margin: 8px 0;
    display: inline-block;
```

```css
    border: 1px solid #ccc;
    border-radius: 4px;
    box-sizing: border-box;
}
input[type=submit]{
    width: 100%;
    background-color: #4CAF50;
    color: white;
    padding: 14px 20px;
    margin: 8px 0;
    border: none;
    border-radius: 4px;
    cursor: pointer;
}
input[type=submit]:hover{
    background-color: #45a049;
}
div{
    border-radius: 5px;
    background-color: #f2f2f2;
    padding: 20px;
}
```

(3) 登录服务 Login.java 文件的代码。

```java
package com.ttt.servlet;
import com.ttt.listener.ThymeleafEngine;
import com.ttt.pojo.Book;
import com.ttt.service.BookService;
import jakarta.servlet.RequestDispatcher;
import jakarta.servlet.ServletException;
import jakarta.servlet.annotation.WebServlet;
import jakarta.servlet.http.HttpServlet;
import jakarta.servlet.http.HttpServletRequest;
import jakarta.servlet.http.HttpServletResponse;
import org.thymeleaf.TemplateEngine;
import org.thymeleaf.context.WebContext;
import org.thymeleaf.web.IWebExchange;
import org.thymeleaf.web.servlet.JakartaServletWebApplication;
import java.io.IOException;
import java.sql.SQLException;
import java.util.List;
@WebServlet("/Login")
public class Login extends HttpServlet {
    @Override
    protected void doPost(HttpServletRequest req, HttpServletResponse resp)
        throws ServletException, IOException{
        req.setCharacterEncoding("UTF-8");
        TemplateEngine templateEngine = (TemplateEngine) getServletContext()
            .getAttribute(ThymeleafEngine.THYMELEAF_ENGINE);
        IWebExchange webExchange = JakartaServletWebApplication
```

```java
            .buildApplication(getServletContext())
            .buildExchange(req, resp);
        WebContext context = new WebContext(webExchange);
        String username = req.getParameter("username");
        String password = req.getParameter("password");
        if((username == null) ||!(username.equalsIgnoreCase("admin")) ||
            (password == null) ||!(password.equalsIgnoreCase("12345"))) {
            context.setVariable("message", "用户名或者密码错误!");
            resp.setContentType("text/html; charset=UTF-8");
            templateEngine.process("login", context, resp.getWriter());
        }
        req.getSession();
        BookService bs = new BookService();
        try{
            List<Book> books = bs.getAllBooks();
            context.setVariable("books", books);
            resp.setContentType("text/html; charset=UTF-8");
            templateEngine.process("books", context, resp.getWriter());
        } catch (SQLException e){
            throw new RuntimeException(e);
        }
    }
}
```

(4) 登录检查过滤器 MyFilter.java 文件的代码。

```java
package com.ttt.filter;
import jakarta.servlet.*;
import jakarta.servlet.annotation.WebFilter;
import jakarta.servlet.http.HttpServletRequest;
import jakarta.servlet.http.HttpServletResponse;
import jakarta.servlet.http.HttpSession;
import java.io.IOException;
@WebFilter(filterName = "MyFilter", urlPatterns = "/BookServlet")
public class MyFilter implements Filter{
    @Override
    public void doFilter(ServletRequest req, ServletResponse resp, FilterChain
      chain) throws IOException, ServletException{
        HttpServletRequest httpServletRequest = (HttpServletRequest)req;
        HttpSession session = httpServletRequest.getSession(false);
        if (session == null){
            String p = httpServletRequest.getContextPath();
            ((HttpServletResponse) resp).sendRedirect(p + "/entry");
            return;
        }
        chain.doFilter(req, resp);
    }
}
```

(5) 数据 POJO 对应的 Book.java 文件的代码。

```java
package com.ttt.pojo;
```

```java
public class Book{
    private Integer id;
    private String name;
    private String publisher;
    private String author;
    private String memo;
    private byte[] cover;
    public Book(Integer id, String name, String publisher, String author,
       String memo, byte[] cover) {
       this.id = id;
       this.name = name;
       this.publisher = publisher;
       this.author = author;
       this.memo = memo;
       this.cover = cover;
    }
    public Integer getId(){
       return id;
    }
    public void setId(Integer id) {
       this.id = id;
    }
    public String getName(){
       return name;
    }
    public void setName(String name){
       this.name = name;
    }
    public String getPublisher(){
       return publisher;
    }
    public void setPublisher(String publisher){
       this.publisher = publisher;
    }
    public String getAuthor(){
       return author;
    }
    public void setAuthor(String author){
       this.author = author;
    }
    public String getMemo() {
       return memo;
    }
    public void setMemo(String memo){
       this.memo = memo;
    }
    public byte[] getCover(){
       return cover;
```

```
    }
    public void setCover(byte[] cover){
        this.cover = cover;
    }
}
```

(6) 主页面 books.html 及其样式文件 books.css。

主页面 books.html 文件的代码如下：

```html
<!DOCTYPE html>
<html xmlns:th="http://www.thymeleaf.org" lang="java">
<head>
  <title>图书信息管理</title>
  <meta http-equiv="Content-Type" content="text/html; charset=UTF-8" />
  <link rel="stylesheet" type="text/css"
        href="../../css/books.css" th:href="@{/css/books.css}"/>
</head>
<body>
  <h2>图形信息管理系统</h2>
  <p><a href="" th:href="@{/BookServlet(f='add')}">新增图书</a></p>
  <div class="book" th:each="book:${books}">
    <img src="" th:src="@{/BookServlet(f='cover',id=${book.id})}" alt=""/>
    <p>书名：[[${book.name}]]</p>
    <p>出版社：[[${book.publisher}]]</p>
    <p>作者：[[${book.author}]]</p>
    <p>简介：[[${book.memo}]]</p>
    <p><a th:href="@{/BookServlet(f='delete',id=${book.id})}">删除</a></p>
    <p><a th:href="@{/BookServlet(f='modify',id=${book.id})}">修改</a></p>
  </div>
</body>
</html>
```

样式文件 books.css 的代码如下：

```css
body{
    width: 1000px;
    margin: 0  auto;
}
h2{
    text-align: center;
}
.book{
    float: left;
    width: 200px;
    height: 400px;
    border: 2px solid rgb(79, 185, 227);
    margin: 10px;
    padding: 10px;
}
img{
    display: block;
    width: 128px;
```

```
        height: 128px;
}
```

(7) 新增图书页面文件 addBook.html 和样式文件 abook.css。

新增图书页面文件 addBook.html 的代码如下：

```
<!DOCTYPE html>
<html xmlns:th="http://www.thymeleaf.org" lang="java">
<head>
  <title>图书信息管理</title>
  <meta http-equiv="Content-Type" content="text/html; charset=UTF-8" />
  <link rel="stylesheet" type="text/css"
       href="../../css/abook.css" th:href="@{/css/abook.css}"/>
</head>
<body>
  <h2>新增/修改图书信息</h2>
  <div>
<form enctype="multipart/form-data" method="POST"
    action="" th:action="@{/BookServlet}">
      <label for="id" hidden="hidden">id</label>
      <input type="text" id="id" name="id" th:value="${book.id}"
          style="display:none">
      <label for="name">图书名</label>
      <input type="text" id="name" name="name"
                th:value="${book.name}" placeholder="图书名称...">
      <label for="publisher">出版社</label>
      <input type="text" id="publisher" name="publisher"
                th:value="${book.publisher}" placeholder="出版社...">
      <label for="author">作者</label>
      <input type="text" id="author" name="author"
                th:value="${book.author}" placeholder="作者...">
      <label for="memo">图书简介</label>
      <input type="text" id="memo" name="memo"
                th:value="${book.memo}" placeholder="图书简介...">
      <label for="cover">头像</label>
      <input type="file" id="cover" name="cover" placeholder="选择图书封面...">
      <input type="submit" value="提交">
    </form>
  </div>
</body>
</html>
```

样式文件 abook.css 的代码如下：

```
input[type=text], input[type=password], input[type=file], p, select{
    width: 100%;
    padding: 12px 20px;
    margin: 8px 0;
    display: inline-block;
    border: 1px solid #ccc;
```

```css
        border-radius: 4px;
        box-sizing: border-box;
}
input[type=submit]{
    width: 100%;
    background-color: #4CAF50;
    color: white;
    padding: 14px 20px;
    margin: 8px 0;
    border: none;
    border-radius: 4px;
    cursor: pointer;
}
input[type=submit]:hover{
    background-color: #45a049;
}
div{
    border-radius: 5px;
    background-color: #f2f2f2;
    padding: 20px;
}
```

（8）图书管理 Servlet 对应的 BookServlet.java 文件的代码。

```java
package com.ttt.servlet;
import com.ttt.listener.ThymeleafEngine;
import com.ttt.pojo.Book;
import com.ttt.service.BookService;
import jakarta.servlet.RequestDispatcher;
import jakarta.servlet.ServletException;
import jakarta.servlet.annotation.MultipartConfig;
import jakarta.servlet.annotation.WebServlet;
import jakarta.servlet.http.HttpServlet;
import jakarta.servlet.http.HttpServletRequest;
import jakarta.servlet.http.HttpServletResponse;
import jakarta.servlet.http.Part;
import org.thymeleaf.TemplateEngine;
import org.thymeleaf.context.WebContext;
importorg.thymeleaf.web.IWebExchange;
import org.thymeleaf.web.servlet.JakartaServletWebApplication;
import java.io.ByteArrayOutputStream;
import java.io.IOException;
import java.io.InputStream;
import java.sql.SQLException;
import java.util.List;
@WebServlet("/BookServlet")
@MultipartConfig
public class BookServlet extends HttpServlet{
    @Override
    protected void doGet(HttpServletRequest req, HttpServletResponse resp)
```

```java
        throws ServletException, IOException{
    req.setCharacterEncoding("UTF-8");
    String f = req.getParameter("f");
    String id = req.getParameter("id");
    TemplateEngine templateEngine = (TemplateEngine) getServletContext()
        .getAttribute(ThymeleafEngine.THYMELEAF_ENGINE);
    IWebExchange webExchange = JakartaServletWebApplication
        .buildApplication(getServletContext())
        .buildExchange(req, resp);
    WebContext context = new WebContext(webExchange);
    BookService bs = new BookService();
    switch(f){
        case "add" ->{
            Book b = new Book(-1, "", "", "", null);
            context.setVariable("book",b);
            resp.setContentType("text/html; charset=UTF-8");
            templateEngine.process("addBook", context, resp.getWriter());
        }
        case "cover" ->{
            try{
                byte[] cover = bs.getBookCoverById(Integer.parseInt(id));
                resp.setContentType("image/jpeg");
                resp.getOutputStream().write(cover);
            } catch (SQLException e){
                throw new RuntimeException(e);
            }
        }
        case "delete" ->{
            try{
                bs.deleteBookById(Integer.parseInt(id));
                List<Book> books = bs.getAllBooks();
                context.setVariable("books", books);
                resp.setContentType("text/html; charset=UTF-8");
                templateEngine.process("books", context, resp.getWriter());
            } catch (SQLException e){
                throw new RuntimeException(e);
            }
        }
        case "modify" ->{
            try{
                Book b = bs.getBookById(Integer.parseInt(id));
                context.setVariable("book", b);
                resp.setContentType("text/html; charset=UTF-8");
                templateEngine.process("addBook", context, resp.getWriter());
            } catch (SQLException e){
                throw new RuntimeException(e);
            }
        }
```

```java
        }
    }
    @Override
    protected void doPost(HttpServletRequest req, HttpServletResponse resp)
        throws ServletException, IOException {
        req.setCharacterEncoding("UTF-8");
        String id = req.getParameter("id");
        String name = req.getParameter("name");
        String publisher = req.getParameter("publisher");
        String author = req.getParameter("author");
        String memo = req.getParameter("memo");
        Part cover = req.getPart("cover");
        byte[] ci = null;
        if (cover != null){
            InputStream is = cover.getInputStream();
            ByteArrayOutputStream baos = new ByteArrayOutputStream();
            byte[] b = new byte[1024];
            while (is.read(b) > 0) {
                baos.write(b);
            }
            ci = baos.toByteArray();
        }
        Book book = null;
        BookService bs = new BookService();
        if ((id != null) && !(id.isBlank()) && (Integer.parseInt(id) != -1)) {
            book = new Book(Integer.parseInt(id), name, publisher, author, memo, ci);
            try{
                bs.updateBook(book);
            } catch (SQLException e){
                throw new RuntimeException(e);
            }
        }
        else{
            book = new Book(-1, name, publisher, author, memo, ci);
            try{
                bs.addBook(book);
            } catch (SQLException e){
                throw new RuntimeException(e);
            }
        }
        List<Book> books = null;
        try{
            books = bs.getAllBooks();
        } catch (SQLException e){
            throw new RuntimeException(e);
        }
        TemplateEngine templateEngine = (TemplateEngine) getServletContext()
            .getAttribute(ThymeleafEngine.THYMELEAF_ENGINE);
```

```
        IWebExchange webExchange = JakartaServletWebApplication
            .buildApplication(getServletContext())
            .buildExchange(req, resp);
        WebContext context = new WebContext(webExchange);
        context.setVariable("books", books);
        resp.setContentType("text/html; charset=UTF-8");
        templateEngine.process("books", context, resp.getWriter());
    }
}
```

(9) 服务及数据操作代码。

图书服务 Service 对应的 BookService.java 文件的代码,图书数据 DAO 对应的 BookDAO.java 文件的代码,数据库连接池对应的 DruidUtil.java 文件的代码及其配置文件 druid.properties,这些代码及文件内容与第 9 章的案例代码一致。

服务及数据库操作代码

运行这个程序,在浏览器地址栏输入/entry,显示如图 10-16 所示的界面。

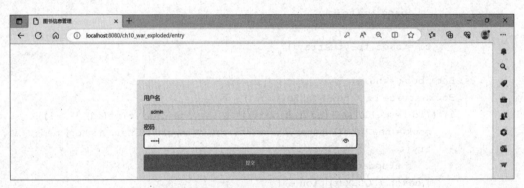

图 10-16　图书信息管理系统登录界面

在图 10-16 的界面中输入 admin 和 12345,显示如图 10-17 所示的主页面。

图 10-17　图形信息管理系统主页面

在图 10-17 所示的主页面中可以新增书籍,也可以修改和删除书籍。

10.8 练习:学生信息管理系统

编写一个学生信息管理系统,具体要求同 9.7 的案例,但是,这里要求使用 Thymeleaf 作为页面表示技术。

第 11 章 JSON、JavaScript 和 Ajax

随着移动设备的出现及其运算能力的提升，人们能够在计算机上处理的业务也能同时在移动设备上进行处理，例如，在智能手机上处理各种业务。一套能够支持多种客户端形式的业务系统是当前典型的业务系统体系结构。在这种支持多种客户端终端形式的系统中，后台服务系统只有一个，只是客户端，也就是前端设备的形式不同而已。这就要求在进行系统设计时，要求后台服务系统能够支持多种形式的前端设备和系统，包括计算机浏览器、安卓 APP、微信小程序等。为了使得一套后台服务系统能够支持多种形式的前端设备，在后台服务系统与前端系统的信息交互上提出了新的要求：后台服务程序与前端程序之间采用 RAW（原始）形式进行数据交互。JSON 正是一种最为常用的 RAW 形式的数据规范。本章对 JSON 和 Ajax 相关技术和应用进行介绍。

11.1 JSON 及其使用

JSON（JavaScript object notation，JavaScript 对象表示法）是目前最为常用的在系统之间进行数据交换的格式规范。

11.1.1 JSON 基础

JSON 是一种轻量级的数据交换格式，它使用"name/value 对"表示数据。JSON 支持两种结构：以"{}"（大括号）表示的对象数据和以"[]"（中括号）表示的数组数据。JSON 表示数据的能力在于它可以使用两种基础格式的组合表示任意复杂的数据。

JSON 基础

11.1.2 为什么需要 JSON

JSON 能够表示任何复杂的数据，同时，JSON 数据又是轻量级和十分简洁的，因此，JSON 广泛地被用于系统之间的数据交互。

设想一种这样的场景：现在需要开发一个智慧校园平台，在这个平台中，学生可以选课、办理学习相关手续、查询考试成绩等；教师可以开设课程、查询班级课表信息、录入学生考试成绩等；管理人员可以查询学生考勤登记、学生学分修满情况、查看学生对老师的教学评价等。这个智慧校园平台不仅要完成如上这些业务功能，还必须支持多种形式的客户端类型，包括基本的计算机浏览器客户端，还可以支持 Android APP、微信小程序等客户端设备。也就是说，后台服务程序必须能够同时支持前端浏览器、Android APP、微信小程序等多种形式,如图 11-1 所示。

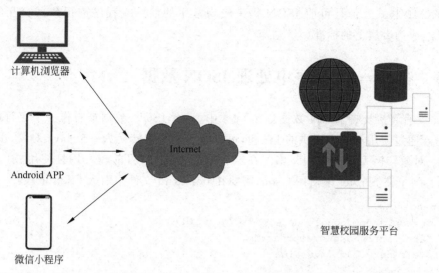

图 11-1　支持多终端类型的智慧校园平台体系结构

由于智慧校园服务平台需要支持多种类型的客户端,因此,对于每个从客户端发送到服务端程序的请求,当服务端程序完成请求处理并形成请求处理结果时,应该将处理结果的原始数据发送到客户端,客户端再根据各自实际情况以适当的形式将结果展示给用户。例如,对于计算机浏览器客户端,则以 HTML 页面的形式将结果展示给用户;对于 Android APP 客户端,则应该以 Android 视图组件的形式将结果展示给用户;对于微信小程序客户端,则应该使用微信的 WXML 组件形式将结果展示给用户。

一般而言,对于支持多类型客户端的服务端程序或平台,只能使用一种原始的数据格式将客户端处理结果返回给客户端,客户端再根据各自的情况采用适当的形式将结果展示给客户。目前,最为常用的从后台服务系统向前端服务程序发送请求处理结果的数据格式就是 JSON 格式,如图 11-2 所示。

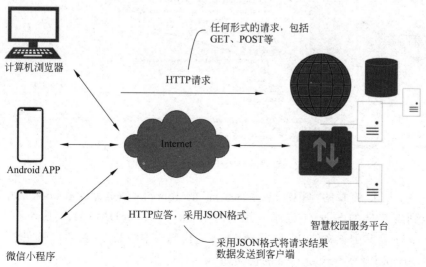

图 11-2　后台服务程序采用 JSON 格式将请求结果发送到客户端

当然,前端客户端程序与后台服务程序之间也可以采用其他形式的数据格式进行数据交互,如 XML 格式。但是目前 JSON 仍是较为常用的前后端程序进行数据交互的格式,JSON 已经成为事实上的标准。

11.1.3 在 Servlet 程序中处理 JSON 数据

Java 没有直接支持 JSON 数据。在 Java 中,一个 JSON 数据被看作一个字符串,称为 JSON 串。通过使用第三方提供的 Jar 包,可以将 JSON 串转换为一个 Java 对象,也可以将一个 Java 对象转换为一个 JSON 串。在这些第三方包中,目前较好用且使用较广泛的是 Gson 包。为了在 Servlet 中使用 Gson,需要在 pom.xml 文件中引入 Gson 依赖:

```xml
<dependency>
    <groupId>com.google.code.gson</groupId>
    <artifactId>gson</artifactId>
    <version>2.10.1</version>
</dependency>
```

下面举例说明如何在 Servlet 程序中将 Java 对象转换为 JSON 串并发送到前端程序。为此,在 IDEA 中新建一个名为 ch11 的 Java Web 程序,删除不必要的其他文件,包括 index.jsp 和 HelloServlet.java,并在其 pom.xml 文件中引入 Gson 依赖。新建完成的 ch11 工程如图 11-3 所示。

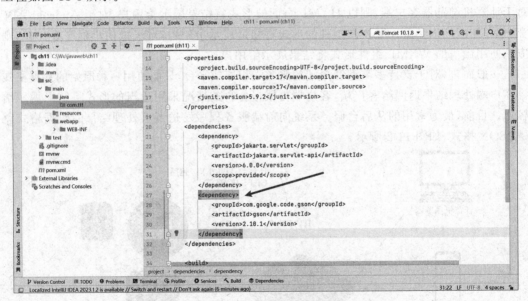

图 11-3 新建的导入了 Gson 依赖的 ch11 工程

编写一个从第 10 章的"图书信息管理系统"的 dbbooks 数据库读取 book 表中的所有书籍并发送到客户端的 Servlet 程序。其中的 Book.java、BooDAO.java、BookService.java、DruidUtil.java 及其配置文件 druid.properties 的代码内容与第 10 章的代码完全一致,此处不再赘述。BookJSONServlet.java 的代码如下:

```java
package com.ttt.servlet;
```

```java
import com.google.gson.Gson;
import com.google.gson.GsonBuilder;
import com.ttt.pojo.Book;
import com.ttt.service.BookService;
import jakarta.servlet.ServletException;
import jakarta.servlet.annotation.WebServlet;
import jakarta.servlet.http.HttpServlet;
import jakarta.servlet.http.HttpServletRequest;
import jakarta.servlet.http.HttpServletResponse;
import java.io.IOException;
import java.io.PrintWriter;
import java.sql.SQLException;
import java.util.List;
@WebServlet("/BookJSONServlet")
public class BookJSONServlet extends HttpServlet{
    @Override
    protected void doGet(HttpServletRequest req, HttpServletResponse resp)
      throws ServletException,IOException {
        req.setCharacterEncoding("UTF-8");
        String f = req.getParameter("f");
        String id = req.getParameter("id");
        BookService bs = new BookService();
        switch(f){
            case "list" ->{
                List<Book> books;
                try{
                    books = bs.getAllBooks();
                } catch (SQLException e){
                    throw new RuntimeException(e);
                }
                resp.setContentType("text/json;charset=utf-8");
                PrintWriter pw = resp.getWriter();
                Gson gson = new GsonBuilder().setDateFormat("yyyy-MM-dd").create();
                pw.print(gson.toJson(books));
            }
            case "cover" ->{
                try {
                    byte[] cover = bs.getBookCoverById(Integer.parseInt(id));
                    resp.setContentType("image/jpeg");
                    resp.getOutputStream().write(cover);
                } catch (SQLException e){
                    throw new RuntimeException(e);
                }
            }
        }
    }
}
```

在这个 Servlet 程序中，根据客户端的请求参数执行不同的行为。当前端的请求参数 f 为 list 时，这个程序从数据库的 book 表中读取所有书籍数据后，然后使用以下代码：

```
Gson gson = new GsonBuilder().setDateFormat("yyyy-MM-dd").create();
pw.print(gson.toJson(books));
```

创建一个 Gson 对象,再使用这个这个 Gson 对象将类型为 List<Book>的对象 books 转换为 JSON 串并发送到前端客户端。运行这个程序,并在浏览器地址栏输入 BookJSONServlet?f=list,将显示如图 11-4 所示的结果。

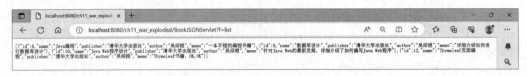

图 11-4 使用 JSON 返回结果数据到客户端浏览器中

从图 11-4 可以看出,后台服务程序返回的请求处理结果是一个 JSON 数组:

[{"id":8,"name":"Java 编程","publisher":"清华大学出版社","author":"吴绍根","memo":"一本不错的编程图书"},{"id":9,"name":"数据库设计","publisher":"清华大学出版社","author":"吴绍根","memo":"详细介绍如何进行数据库设计"},{"id":10,"name":"Java Web 程序设计","publisher":"清华大学出版社","author":"吴绍根","memo":"针对 Java Web 的最新发展,详细介绍了如何编写 Java Web 程序"},{"id":12,"name":"Thymeleaf 页面编程","publisher":"清华大学出版社","author":"吴绍根","memo":"Thymeleaf 书籍,OK,OK"}]

为了便于观察,将这个结果稍作整理,结果如下:

```
[
    {
        "id":8,
        "name":"Java 编程",
        "publisher":"清华大学出版社",
        "author":"吴绍根",
        "memo":"一本不错的编程图书"
    },
    {
        "id":9,
        "name":"数据库设计",
        "publisher":"清华大学出版社",
        "author":"吴绍根",
        "memo":"详细介绍如何进行数据库设计"
    },
    {
        "id":10,
        "name":"Java Web 程序设计",
        "publisher":"清华大学出版社",
        "author":"吴绍根",
        "memo":"针对 Java Web 的最新发展,详细介绍了如何编写 Java Web 程序"
    },
    {
        "id":12,
        "name":"Thymeleaf 页面编程",
        "publisher":"清华大学出版社",
        "author":"吴绍根",
```

```
    "memo":"Thymeleaf 图书,OK,OK"
  }
]
```

对于浏览器的请求"BookJSONServlet?f=list",后台服务程序以 JSON 格式返回了所有书籍的信息。如果 Android APP 中发送同样的请求,那么 APP 程序也将收到同样的 JSON 结果;如果微信小程序也发送了同样的请求,那么微信小程序也将收到同样的 JSON 结果。也就是说,对于 BookJSONServlet 程序而言,它不管请求来自哪里及来自哪种形式的客户端,同样的请求将得到同样的数据结果。对于不同的客户端形式,各自根据自己的特点将得到的 JSON 数据展示给用户。这就是使用 JSON 的好处:返回数据与客户端形式无关。

11.2 JavaScript 和 Ajax

为了在浏览器中显示后台服务程序返回的 JSON 数据结果,前端程序需要对结果数据进行再次处理,以便以页面形式在浏览器中显示出来。这就需要在页面 HTML 代码中使用 JavaScript 对数据进行处理,形成 HTML 页面标签,进而以友好形式给用户展示数据结果。就从访问上一节的"BookJSONServlet?f=list"并将数据以友好的 HTML 页面形式展示开始。

11.2.1 展示所有书籍信息

在 ch11 工程中新建名为 books.html 的页面,用以显示从后台服务程序返回的书籍列表信息。books.html 的代码如下:

```
<!DOCTYPE html>
<html lang="en">
<head>
    <title>图书信息管理</title>
    <meta http-equiv="Content-Type" content="text/html; charset=UTF-8" />
    <link rel="stylesheet" type="text/css" href="../css/books.css">
    <script type="text/javascript" src="../js/books.js"></script>
</head>
<body onload="loadBooks()">
    <h2>图形信息管理</h2>
    <div id="books">
    </div>
</body>
</html>
```

books.html 的代码非常简单,只是用<h2>标签显示了一段文字。那么,它是如何加载所有书籍信息的呢?注意代码中粗体字的语句,这条语句告知浏览器在加载页面时要执行名称为 loadBooks() 的 JavaScript 函数。这个函数在文件"../js/books.js"中定义,并且通过以下语句导入到 HTML 页面中:

253

```html
<script type="text/javascript" src="../js/books.js"></script>
```

books.js 的代码如下:

```javascript
function loadBooks(){
    let xrq = new XMLHttpRequest();
    xrq.onreadystatechange = function(){
        let bks;
        if (xrq.readyState === XMLHttpRequest.DONE){
            if (xrq.status === 200){
                bks = JSON.parse(xrq.responseText);
                handleBooks(bks);
            } else{
                alert("请求遇到了问题!");
            }
        }
    };
    xrq.open("GET", "../BookJSONServlet?f=list");
    xrq.send();
}
function handleBooks(bks){
    let books = document.getElementById("books");
    for (let i=0; i<bks.length; i++){
        let div = document.createElement('div');
        div.className = 'book';
        div.innerHTML = '<img src="../BookJSONServlet?f=cover&id=' + bks[i].
                id + '"' + ' alt=""/>' +
                '<p>书名: ' + bks[i].name + '</p>' +
                '<p>出版社: ' + bks[i].publisher + '</p>' +
                '<p>书名: ' + bks[i].author + '</p>' +
                '<p>书名: ' + bks[i].memo + '</p>';
        books.appendChild(div);
    }
}
```

在 loadBooks() 函数中,使用如下语句并采用 Ajax 向后台服务程序发送请求。

```javascript
let xrq = new XMLHttpRequest();
```

和

```javascript
xrq.open("GET", "../BookJSONServlet?f=list");
xrq.send();
```

同时,采用如下语句对后台返回的数据使用 JavaScript 进行处理,并动态插入页面的 DOM 页面中:

```javascript
let books = document.getElementById("books");
```

和

```javascript
books.appendChild(div);
```

为了按要求显示书籍信息,还定义了 books.css 样式文件,books.css 文件内容如下。

```css
body{
```

```
        width: 1000px;
        margin: 0   auto;
    }
    h2{
        text-align: center;
    }
    .book{
        float: left;
        width: 200px;
        height: 400px;
        border: 2px solid rgb(79, 185, 227);
        margin: 10px;
        padding: 10px;
    }
    img{
        display: block;
        width: 128px;
        height: 128px;
    }
```

现在运行这个程序，在浏览器地址栏输入 html/books.html，显示如图 11-5 所示的页面。

图 11-5　采用 Ajax 读取书籍列表信息

从图 11-5 可以看出，books.html 页面显示了全部书籍信息，这正是使用 Ajax 动态从后台服务程序读取书籍信息并使用 JavaScript 将书籍信息显示在浏览器上的效果。

11.2.2　Ajax

Ajax 是一种在 JavaScript 代码中动态向后台服务程序发送 HTTP 请求，获取后台服务程序响应结果的一种技术方式。因此，要使用 Ajax，必须了解 JavaScript 语言及使用 JavaScript 程序操作 HTML DOM 文档。关于 JavaScript 语言的相关内容已经超出了本书范畴，不在此介绍 JavaScript 语言及其使用。

1. Ajax 概述

Ajax 是 Asynchronous JavaScript and XML 的缩写。注意,虽然 Ajax 的最后一个单词是 XML,但是需要强调的是,Ajax 与 XML 没有关系。简单来说,Ajax 就是在 JavaScript 程序中使用浏览器内建的 XMLHttpRequest 对象,向后台服务程序发送 HTTP 请求并接收后台服务程序返回的 HTPP 响应的一种技术。Ajax 工作的一般过程如图 11-6 所示。

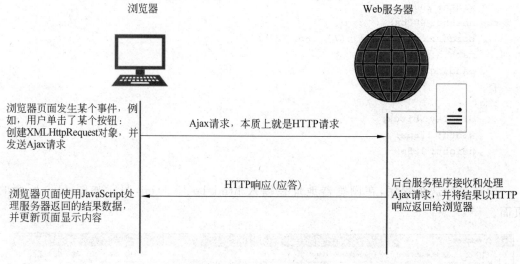

图 11-6　Ajax 的工作过程

从图 11-6 可以看出,在页面程序中使用 Ajax 的一般过程如下:网页中发生一个事件(页面加载、用户单击某个按钮等)→由 JavaScript 代码创建 XMLHttpRequest 对象→使用创建的 XMLHttpRequest 对象向 Web 服务器发送 HTTP 请求→服务器处理该请求→服务器将响应发送回网页→页面 JavaScript 程序读取响应→页面 JavaScript 代码执行正确的动作,如将数据显示在更新的页面中等。

2. XMLHttpRequest 对象及其属性和方法

所有现代的浏览器都支持 XMLHttpRequest 对象,也就是说,在页面的 JavaScript 代码中可以使用如下语句创建 XMLHttpRequest 对象:

```
let xhttp = new XMLHttpRequest();
```

一旦创建了 XMLHttpRequest 对象,可以设置该对象的相关属性,然后使用其 send() 方法即可发送 HTTP 请求。XMLHttpRequest 对象的属性及其含义如表 11-1 所示。

表 11-1　XMLHttpRequest 对象的属性及其含义

序号	属性名称	含　义
1	onload	这是一个回调函数,定义当浏览器接收到后台服务程序返回的处理结果时要调用的函数
2	onreadystatechange	这是一个回调函数,定义当 readyState 属性发生变化时调用的函数

续表

序号	属性名称	含义
3	readyState	保存 XMLHttpRequest 的状态,可能的取值及其含义如下:0 表示请求未初始化,1 表示服务器连接已建立,2 表示请求已收到,3 表示正在处理请求,4 表示请求已完成且响应已就绪
4	responseText	后台服务程序返回的是以字符串形式返回响应数据
5	responseXML	后台服务程序返回的是以 XML 数据返回响应数据
6	status	后台服务程序返回请求处理结果的状态码,例如:200 代表 OK,403 代表 Forbidden,404 代表 Not Found。详细的状态码可参见第 2 章对 HTTP 的介绍
7	statusText	返回状态码对应的文本解释,例如,状态码为 200 时,对应的文本解释为 OK;状态码为 404 时,对应的文本解释为 Not Found 等

XMLHttpRequest 对象除了包含表 11-1 所定义的属性外,还包含一些方法。XMLHttpRequest 对象的方法及其含义如表 11-2 所示。

表 11-2　XMLHttpRequest 对象的方法及其含义

序号	方法名称	含义
1	open(method, url, async, user, psw)	构建 Ajax 请求类型及基本参数。method 表示请求类型为 GET 或 POST,url 表示服务端位置,async 表示 true(异步)或 false(同步),user 表示可选的用户名,psw 表示可选的密码
2	send()	向服务器发送请求,用于 GET 请求
3	send(string params)	向服务器发送请求,用于 POST 请求
4	setRequestHeader(string name, string value)	将名/值对添加到要发送的请求头
5	getResponseHeader(string name)	返回特定的响应头部信息
6	getAllResponseHeaders()	返回所有响应头部信息
7	abort()	取消当前请求

3. 再次观察获取和显示所有书籍信息的 books.js 代码

现在,再回头看看 11.2.1 小节中用于获取并显示所有书籍信息的 JavaScript 代码,也就是 books.js 的代码。books.js 的代码如下:

```
function loadBooks(){
    let xrq = new XMLHttpRequest();
    xrq.onreadystatechange = function(){
        let bks;
        if (xrq.readyState === XMLHttpRequest.DONE){
            if (xrq.status === 200) {
                bks = JSON.parse(xrq.responseText);
                handleBooks(bks);
            } else{
                alert("请求遇到了问题!");
            }
```

```
        }
    };
    xrq.open("GET", "../BookJSONServlet?f=list");
    xrq.send();
}
function handleBooks(bks){
    let books = document.getElementById("books");
    for (let i=0; i<bks.length; i++){
        let div = document.createElement('div');
        div.className = 'book';
        div.innerHTML = '<img src="../BookJSONServlet?f=cover&id=' + bks[i].
        id + '"' + ' alt=""/>' + '<p>书名: ' + bks[i].name + '</p>' + '<p>出版
        社: ' + bks[i].publisher + '</p>' + '<p>书名: ' + bks[i].author + '</p>' +
        '<p>书名: ' + bks[i].memo + '</p>';
        books.appendChild(div);
    }
}
```

在 loadBooks()函数中,首先使用语句"let xrq = new XMLHttpRequest();"创建了 XMLHttpRequest 对象。然后,通过如下语句:

```
xrq.onreadystatechange = function(){
    let bks;
    if (xrq.readyState === XMLHttpRequest.DONE){
        if (xrq.status === 200){
            bks = JSON.parse(xrq.responseText);
            handleBooks(bks);
        } else{
            alert("请求遇到了问题!");
        }
    }
};
```

设置当 xrq 的 readyState 状态属性发生变化时要调用的回调函数。在这个回调函数中,首先检查 xrq 的 readyState 的值是否为 XMLHttpRequest.DONE,也就是 4。如果是 4,则表示请求已经被后台服务程序完成处理,然会再次检查 xrq 的状态码,也就是 xrq.status 属性值是否为 200,即 OK,表示后台服务程序正确地处理了请求并返回处理结果。如果是,则调用语句"bks = JSON.parse(xrq.responseText);"将服务返回的 JSON 字符串转换为 JavaScript 的对象,然后调用"handleBooks(bks);"对结果数据进行处理。

大家还记得 URL 为"../BookJSONServlet?f=list"的后台服务程序返回的结果数据及其格式吗?为了便于阅读,再次给出示例结果数据如下:

```
[
{
    "id":8,
    "name":"Java 编程",
    "publisher":"清华大学出版社",
    "author":"吴绍根",
    "memo":"一本不错的编程图书"
},
```

```
    {
        "id":9,
        "name":"数据库设计",
        "publisher":"清华大学出版社",
        "author":"吴绍根",
        "memo":"详细介绍如何进行数据库设计"
    },
    {
        "id":10,
        "name":"Java Web 程序设计",
        "publisher":"清华大学出版社",
        "author":"吴绍根",
        "memo":"针对 Java Web 的最新发展,详细介绍了如何编写 Java Web 程序"
    },
    {
        "id":12,
        "name":"Thymeleaf 页面编程",
        "publisher":"清华大学出版社",
        "author":"吴绍根",
        "memo":"Thymeleaf 图书,OK,OK"
    }
]
```

结果数据是一个 JSON 数组。handleBooks()函数就是要处理这个 JSON 数组并将每本书籍信息显示在页面中。在 handleBooks()函数中,首先使用语句"let books = document.getElementById("books");"获得 books.html 页面中的 id 为 books 的<div>标签的元素。然后,使用如下语句:

```
for (let i=0; i<bks.length; i++){
    let div = document.createElement('div');
    div.className = 'book';
    div.innerHTML = '<img src="../BookJSONServlet?f=cover&id=' + bks[i].id + '"' + ' alt=""/>' + '<p>书名:' + bks[i].name + '</p>' + '<p>出版社:' + bks[i].publisher + '</p>' + '<p>书名:' + bks[i].author + '</p>' + '<p>书名:' + bks[i].memo + '</p>';
    books.appendChild(div);
}
```

对每一本书籍信息进行处理:为每一本书籍创建一个用以显示书籍信息的<div>标签;然后将书籍信息显示在这个<div>标签下;最后,将这个标签添加到 id 为 books 的<div>下,其作用相当于将每本书籍信息显示在 id 为 books 的<div>下。

JSP vs Thymeleaf vs JavaScript

11.3 案例：图书信息管理系统

作为对 Servlet、JSON、Ajax 等技术的综合使用，本案例设计一个图书信息管理系统，用于对图书信息进行管理，具体要求同 9.6 节的案例。但是，这里要求使用 Javascript 和 Ajax 请求后台服务程序数据并使用 JavaScript 操作 HTML 页面的 DOM 文档进行图书信息展示。

11.3.1 案例目标

编写一个简单的图书管理系统，图书信息保存在数据库中。每本图书包含如下基本信息：名称、出版社、作者、书籍简介、书籍封面图片。在主页中，显示已经录入的所有图书信息，对已有的图书可以进行编辑，包括删除及修改图书信息。同时，在主页面中还包括一个"新增"按钮，单击该按钮，可以新增图书到数据库中。

11.3.2 案例分析

在对图书信息进行管理之前，需要先登录到系统中，并且对于未登录的访问，系统将自动将访问引导到登录页面。因此，需要设计和编写如下程序文件。

（1）登录页面：用 login.html 文件，还有样式文件 login.css 及 login.js 文件。

（2）登录服务 Servlet：用 Login.java 文件，完成登录验证并创建 Session。登录成功后直接跳转到主页面，登录失败则再次显示登录页面。

（3）登录检查过滤器：用 MyFilter.java 文件，放行登录用户的访问，将未登录用户直接引导到登录页面。

（4）数据 POJO：用 Book.java 文件，表示图书书籍类，是一个简单的 POJO 类。

（5）主页面：用 booksN.html 文件，还有样式文件 booksN.css 及其 booksN.js 文件。在这个主页面上显示所有已有图书信息，用户可以删除、修改或新增图书信息。

（6）新增图书页面：用 addBook.html 文件，用户在这个页面可以录入新的图书信息。

（7）图书管理 Servlet：用 BookServlet.java 文件，完成图书的增、删、改参数的接收，并将处理结果返回给客户端。

（8）图书服务 Service：用 BookService.java 文件，完成具体图书的增、删、改服务。它被 BookServlet 使用，完成具体的业务处理。

（9）图书数据 DAO：用 BookDAO.java 文件，完成具体数据信息的数据库操作。

（10）数据库连接池：用 DruidUtil.java 文件及其配置文件 druid.properties，完成数据库连接池的管理。

11.3.3 案例实施

首先创建数据库及相关数据表。在 MySQL 下创建 dbbooks 数据库和 book 数据表，相

关语句如下:

```
CREATE SCHEMA 'dbbooks' DEFAULT CHARACTER SET utf8mb4;
CREATE TABLE 'dbbooks'.'book' (
  'id' INT NOT NULL AUTO_INCREMENT,
  'name' VARCHAR(45) NOT NULL,
  'publisher' VARCHAR(200) NOT NULL,
  'author' VARCHAR(45) NOT NULL,
  'memo' TEXT(2000) NULL,
  'cover' LONGBLOB NULL,
  PRIMARY KEY ('id'));
```

(1) 登录页面。登录页面包括的文件有 login.html、login.css 及 login.js。在登录页面中,用户输入用户名、密码后,单击"提交"按钮。登录界面并不是直接调用<form>表单的提交功能将数据提交给后台服务程序处理,而是使用 JavaScript 代码,通过 Ajax 方式将登录数据提交给后台处理。login.html 代码如下:

```html
<!DOCTYPE html>
<html lang="java">
<head>
    <title>图书信息管理</title>
    <meta http-equiv="Content-Type" content="text/html; charset=UTF-8" />
    <link rel="stylesheet" type="text/css" href="../css/login.css">
    <script type="text/javascript" src="../js/login.js"></script>
</head>
<body>
<div>
    <label for="username">用户名</label>
    <input type="text" id="username" name="username" placeholder="用户名...">
    <label for="password">密码</label>
    <input type="password" id="password" name="password" placeholder="密码...">
    <input type="submit" value="提交" onclick="login()">
</div>
</body>
</html>
```

涉及的样式文件 login.css 的内容如下:

```css
div{
    width: 600px;
    margin: 100px auto;
}
input[type=text], input[type=password]{
    width: 100%;
    padding: 12px 20px;
    margin: 8px 0;
    display: inline-block;
    border: 1px solid #ccc;
    border-radius: 4px;
```

```css
    box-sizing: border-box;
}
input[type=submit]{
    width: 100%;
    background-color: #4CAF50;
    color: white;
    padding: 14px 20px;
    margin: 8px 0;
    border: none;
    border-radius: 4px;
    cursor: pointer;
}
input[type=submit]:hover{
    background-color: #45a049;
}
div{
    border-radius: 5px;
    background-color: #f2f2f2;
    padding: 20px;
}
```

用于发送登录 Ajax 请求的 JavaScript 代码 login.js 如下：

```javascript
function login(){
    let xrq = new XMLHttpRequest();
    xrq.onreadystatechange = function(){
        if (xrq.readyState === XMLHttpRequest.DONE){
            if (xrq.status === 200){
                let result = JSON.parse(xrq.responseText);
                if (result.success === true){
                    window.open("./booksN.html", "_self");
                }
                else{
                    alert(result.message +",请重新登录!");
                    document.getElementById("username").value = "";
                    document.getElementById("password").value = "";
                }
            } else{
                alert("请求遇到了问题!");
            }
        }
    };
    xrq.open("POST", "../Login");
    xrq.setRequestHeader("Content-type", "application/x-www-form-
        urlencoded");
    let username = document.getElementById("username").value;
    let password = document.getElementById("password").value;
    xrq.send("username=" + username + "&password=" + password);
}
```

（2）登录服务 Servlet。对应文件为 Login.java。完成登录验证并创建 Session。登录成功后直接跳转到主页面，登录失败则再次显示登录。Login.java 代码如下：

```java
package com.ttt.servlet;
import com.google.gson.Gson;
import com.google.gson.GsonBuilder;
import jakarta.servlet.ServletException;
import jakarta.servlet.annotation.WebServlet;
import jakarta.servlet.http.HttpServlet;
import jakarta.servlet.http.HttpServletRequest;
import jakarta.servlet.http.HttpServletResponse;
import java.io.IOException;
import java.io.PrintWriter;
@WebServlet("/Login")
public class Login extends HttpServlet{
    @Override
    protected void doPost(HttpServletRequest req, HttpServletResponse resp)
      throws ServletException, IOException{
        req.setCharacterEncoding("UTF-8");
        Gson gson = new GsonBuilder().setDateFormat("yyyy-MM-dd").create();
        String username = req.getParameter("username");
        String password = req.getParameter("password");
        if((username == null) || !(username.equalsIgnoreCase("admin")) ||
           (password == null) || !(password.equalsIgnoreCase("12345"))) {
            Result rt = new Result();
            rt.success = false;
            rt.message = "登录失败";
            resp.setContentType("application/json;charset=utf-8");
            PrintWriter pw = resp.getWriter();
            pw.print(gson.toJson(rt));
            return;
        }
        req.getSession();
        Result rt = new Result();
        rt.success = true;
        rt.message = "登录成功";
        resp.setContentType("application/json;charset=utf-8");
        PrintWriter pw = resp.getWriter();
        pw.print(gson.toJson(rt));
    }
    private static class Result{
        private boolean success;
        private String message;
    }
}
```

（3）登录检查过滤器。对应文件为 MyFilter.java，用于放行登录用户的访问，对未登录用户直接引导到登录页面。MyFilter.java 代码如下：

```java
package com.ttt.filter;
import jakarta.servlet.*;
import jakarta.servlet.annotation.WebFilter;
import jakarta.servlet.http.HttpServletRequest;
import jakarta.servlet.http.HttpServletResponse;
import jakarta.servlet.http.HttpSession;
import java.io.IOException;
@WebFilter(filterName = "MyFilter", urlPatterns = {"/BookServlet", "/html/booksN.html", "/html/addBook.html"})
public class MyFilter implements Filter{
    @Override
    public void doFilter(ServletRequest req, ServletResponse resp, FilterChain chain)
      throws IOException, ServletException {
        HttpServletRequest httpServletRequest = (HttpServletRequest)req;
        HttpSession session = httpServletRequest.getSession(false);
        if (session == null){
            String p = httpServletRequest.getContextPath();
            ((HttpServletResponse) resp).sendRedirect(p + "/html/login.html");
            return;
        }
        chain.doFilter(req, resp);
    }
}
```

（4）数据 POJO。对应文件为 Book.java，是图书书籍类。Book.java 的代码如下：

```java
package com.ttt.pojo;
public class Book{
    private Integer id;
    private String name;
    private String publisher;
    private String author;
    private String memo;
    private byte[] cover;
    public Book(Integer id, String name, String publisher, String author,
      String memo, byte[] cover) {
        this.id = id;
        this.name = name;
        this.publisher = publisher;
        this.author = author;
        this.memo = memo;
        this.cover = cover;
    }
    public Integer getId(){
        return id;
    }
    public void setId(Integer id){
        this.id = id;
    }
```

```java
    public String getName(){
        return name;
    }
    public void setName(String name){
        this.name = name;
    }
    public String getPublisher(){
        return publisher;
    }
    public void setPublisher(String publisher){
        this.publisher = publisher;
    }
    public String getAuthor(){
        return author;
    }
    public void setAuthor(String author){
        this.author = author;
    }
    public String getMemo(){
        return memo;
    }
    public void setMemo(String memo){
        this.memo = memo;
    }
    public byte[] getCover(){
        return cover;
    }
    public void setCover(byte[] cover){
        this.cover = cover;
    }
}
```

(5) 主页面。主页面包括 booksN.html、booksN.css 及 booksN.js 文件。booksN.html 文件的代码如下：

```html
<!DOCTYPE html>
<html lang="en">
<head>
    <title>图书信息管理</title>
    <meta http-equiv="Content-Type" content="text/html; charset=UTF-8" />
    <link rel="stylesheet" type="text/css" href="../css/booksN.css"/>
    <script type="text/javascript" src="../js/booksN.js"></script>
</head>
<body onload="loadBooks()">
    <h2>图书信息管理</h2>
    <p><a href="./addBook.html">新增图书</a></p>
    <div id="books">
    </div>
</body>
```

```
</html>
```

booksN.css 文件的代码如下：

```css
body{
    width: 1000px;
    margin: 0  auto;
}
h2{
    text-align: center;
}
.book{
    float: left;
    width: 200px;
    height: 400px;
    border: 2px solid rgb(79, 185, 227);
    margin: 10px;
    padding: 10px;
}
img{
    display: block;
    width: 128px;
    height: 128px;
}
```

booksN.js 文件的代码如下：

```javascript
function loadBooks(){
    let xrq = new XMLHttpRequest();
    xrq.onreadystatechange = function(){
        let bks;
        if (xrq.readyState === XMLHttpRequest.DONE){
            if (xrq.status === 200){
                bks = JSON.parse(xrq.responseText);
                handleBooks(bks);
            } else{
                alert("请求遇到了问题!");
            }
        }
    };
    xrq.open("GET", "../BookServlet?f=list");
    xrq.send();
}
function handleBooks(bks){
    let books = document.getElementById("books");
    for (let i=0; i<bks.length; i++){
        let div = document.createElement('div');
        div.className = 'book';
        div.innerHTML = '<img src="../BookServlet?f=cover&id=' + bks[i].id + '"' +
        ' alt=""/>' + '<p>书名：' + bks[i].name + '</p>' + '<p>出版社：' + bks
        [i].publisher + '</p>' + '<p>书名：' + bks[i].author + '</p>' + '<p>书名：
```

```
                     ' + bks[i].memo + '</p>' + '<p><a href="javascript:deleteBook(' + bks
                     [i].id + ')"' + '>删除</a></p>' + '<p><a href="javascript:modifyBook('
                     + bks[i].id + ')"' + '>修改</a></p>';
                books.appendChild(div);
        }
}
function deleteBook(id){
        let xrq = new XMLHttpRequest();
        xrq.onreadystatechange = function(){
                let bks;
                if (xrq.readyState === XMLHttpRequest.DONE) {
                        if (xrq.status === 200){
                                let result = JSON.parse(xrq.responseText);
                                if (result.success === true){
                                        window.open("./booksN.html", "_self");
                                }
                                else {
                                        alert(result.message);
                                }
                        } else{
                                alert("请求遇到了问题!");
                        }
                }
        };
        xrq.open("GET", "../BookServlet?f=delete&id=" + id);
        xrq.send();
}
function modifyBook(id){
        alert(id);
}
```

（6）新增图书页面。新增图书页面包括 addBook.html、abook.css 和 abook.js 等文件。addBook.html 文件的代码如下：

```
<!DOCTYPE html>
<html lang="java">
<head>
  <title>图书信息管理</title>
  <meta http-equiv="Content-Type" content="text/html; charset=UTF-8" />
  <link rel="stylesheet" type="text/css" href="../css/abook.css"/>
  <script type="text/javascript" src="../js/abook.js"></script>
</head>
<body>
  <h2>新增/修改图书信息</h2>
  <div>
    <form id="form" enctype="multipart/form-data" method="POST">
      <label for="id" hidden="hidden">id</label>
      <input type="text" id="id" name="id" style="display:none">
      <label for="name">书籍名</label>
```

```html
            <input type="text" id="name" name="name" placeholder="书籍名称...">
            <label for="publisher">出版社</label>
            <input type="text" id="publisher" name="publisher" placeholder="出版社...">
            <label for="author">作者</label>
            <input type="text" id="author" name="author" placeholder="作者...">
            <label for="memo">图书简介</label>
            <input type="text" id="memo" name="memo" placeholder="图书简介...">
            <label for="cover">头像</label>
            <input type="file" id="cover" name="cover" placeholder="选择图书封面...">
            <input type="submit" value="提交" onclick="smform();return false;">
        </form>
    </div>
</body>
</html>
```

abook.css 样式文件的代码如下：

```css
input[type=text], input[type=password], input[type=file], p, select{
    width: 100%;
    padding: 12px 20px;
    margin: 8px 0;
    display: inline-block;
    border: 1px solid #ccc;
    border-radius: 4px;
    box-sizing: border-box;
}
input[type=submit]{
    width: 100%;
    background-color: #4CAF50;
    color: white;
    padding: 14px 20px;
    margin: 8px 0;
    border: none;
    border-radius: 4px;
    cursor: pointer;
}
input[type=submit]:hover{
    background-color: #45a049;
}
div{
    border-radius: 5px;
    background-color: #f2f2f2;
    padding: 20px;
}
```

abook.js 文件的代码如下：

```js
function smform(){
    let f = document.getElementById("form");
```

```javascript
        let formData = new FormData(f);
        let xrq = new XMLHttpRequest();
        xrq.onreadystatechange = function(){
            if (xrq.readyState === XMLHttpRequest.DONE){
                if (xrq.status === 200){
                    let result = JSON.parse(xrq.responseText);
                    if (result.success === true){
                        window.open("./booksN.html", "_self");
                    }
                    else{
                        alert(result.message);
                    }
                } else{
                    alert("请求遇到了问题!");
                }
            }
        };
        xrq.open("POST", "../BookServlet");
        xrq.send(formData);
    }
```

（7）图书管理 Servlet。对应的文件为 BookServlet.java。代码如下：

```java
package com.ttt.servlet;
import com.google.gson.Gson;
import com.google.gson.GsonBuilder;
import com.ttt.pojo.Book;
import com.ttt.service.BookService;
import jakarta.servlet.ServletException;
import jakarta.servlet.annotation.MultipartConfig;
import jakarta.servlet.annotation.WebServlet;
import jakarta.servlet.http.HttpServlet;
import jakarta.servlet.http.HttpServletRequest;
import jakarta.servlet.http.HttpServletResponse;
import jakarta.servlet.http.Part;
import java.io.ByteArrayOutputStream;
import java.io.IOException;
import java.io.InputStream;
import java.io.PrintWriter;
import java.sql.SQLException;
import java.util.List;
@WebServlet("/BookServlet")
@MultipartConfig
public class BookServlet extends HttpServlet{
    @Override
    protected void doGet(HttpServletRequest req, HttpServletResponse resp)
      throws ServletException, IOException {
        req.setCharacterEncoding("UTF-8");
        String f = req.getParameter("f");
```

```java
String id = req.getParameter("id");
BookService bs = new BookService();
switch(f){
    case "list" ->{
        List<Book> books;
        try{
            books = bs.getAllBooks();
        } catch (SQLException e) {
            throw new RuntimeException(e);
        }
        resp.setContentType("text/json;charset=utf-8");
        PrintWriter pw = resp.getWriter();
        Gson gson = new GsonBuilder().setDateFormat("yyyy-MM-dd").create();
        pw.print(gson.toJson(books));
    }
    case "cover" ->{
    try {
        byte[] cover = bs.getBookCoverById(Integer.parseInt(id));
        resp.setContentType("image/jpeg");
        resp.getOutputStream().write(cover);
    } catch (SQLException e) {
        throw new RuntimeException(e);
    }
    }
    case "delete" ->{
        try{
            int count = bs.deleteBookById(Integer.parseInt(id));
            Gson gson = new GsonBuilder().setDateFormat("yyyy-MM-dd").
            create();
            Result rt = new Result();
            if (count >= 1){
                rt.success = true;
                rt.message = "删除成功";
            }
            else {
                rt.success = false;
                rt.message = "删除失败";
            }
            resp.setContentType("application/json;charset=utf-8");
            PrintWriter pw = resp.getWriter();
            pw.print(gson.toJson(rt));
        } catch (SQLException e) {
            throw new RuntimeException(e);
        }
    }
    case "modify" ->{
        try{
            Book b = bs.getBookById(Integer.parseInt(id));
```

```java
            } catch (SQLException e) {
                throw new RuntimeException(e);
            }
        }
    }
}
@Override
protected void doPost(HttpServletRequest req, HttpServletResponse resp)
    throws ServletException, IOException{
    req.setCharacterEncoding("UTF-8");
    String id = req.getParameter("id");
    String name = req.getParameter("name");
    String publisher = req.getParameter("publisher");
    String author = req.getParameter("author");
    String memo = req.getParameter("memo");
    Part cover = req.getPart("cover");
    byte[] ci = null;
    if (cover !=null) {
        InputStream is = cover.getInputStream();
        ByteArrayOutputStream baos = new ByteArrayOutputStream();
        byte[] b = new byte[1024];
        while (is.read(b) > 0) {
            baos.write(b);
        }
        ci = baos.toByteArray();
    }
    Book book = null;
    BookService bs = new BookService();
    if ((id != null) && !(id.isBlank()) && (Integer.parseInt(id) != -1)) {
        book = new Book(Integer.parseInt(id), name, publisher, author, memo, ci);
        try{
            bs.updateBook(book);
        } catch (SQLException e) {
            throw new RuntimeException(e);
        }
    }
    else{
        book = new Book(-1, name, publisher, author, memo, ci);
        try{
            Book b = bs.addBook(book);
        } catch (SQLException e) {
            throw new RuntimeException(e);
        }
    }
    Gson gson = new GsonBuilder().setDateFormat("yyyy-MM-dd").create();
    Result rt = new Result();
    rt.success = true;
    rt.message = "操作成功";
    resp.setContentType("application/json;charset=utf-8");
```

```
        PrintWriter pw = resp.getWriter();
        pw.print(gson.toJson(rt));
    }
    private static class Result {
        private boolean success;
        private String message;
    }
}
```

（8）服务及数据库操作代码。图书服务 Service 对应文件为 BookService.java，图书数据 DAO 对应文件为 BookDAO.java 代码；数据库连接池对应文件为 DruidUtil.java，其配置文件对应文件为 druid.properties，这些文件内容与第 9 章案例的代码一致。

服务及数据库操作代码

运行这个程序，在浏览器地址栏输入"/html/login.html"，显示如图 11-7 的登录页面。

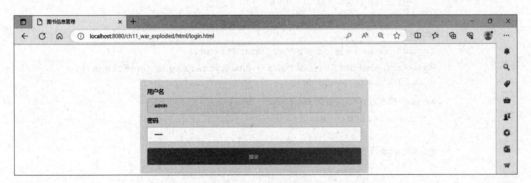

图 11-7　登录页面

在图 11-7 的界面中输入用户名 admin 和密码 12345，单击"提交"按钮，将显示如图 11-8 所示的系统主页面。

图 11-8　图书管理主页面

在图 11-8 的界面中,可以新增、删除和修改图书信息。

11.4 练习:完善图书信息管理系统

在 11.3 节的图书信息管理系统中,案例程序没有完成对图书的修改功能。仔细阅读和理解 11.3 节的案例程序,完成对图书信息的修改功能。

第 3 部分

Java Web 高级特性

第 12 章 Servlet 高级技术

第 12 章 Servlet 高级技术

Java Web 作为 Java EE 的重要组成部分,与其相关的规范和技术标准一直在不断发展。目前 Jakarta EE 的最新版本已经发展到 10.0 版本,与之相应的 Servlet 技术规范也升级到了 6.0 版本。对于 Java Web 而言,Servlet 是其核心和关键技术。从 Servlet 3.0 开始,在 Servlet 的技术规范中引入了新的技术特性,其中的 AsyncContext 技术和 Non Blocking I/O 技术就是其中的重要部分。

12.1 AsyncContext 异步处理请求技术

在第 3 章中详细介绍了 Servlet 相关技术及其应用。现在回顾一下 Servlet 处理前端请求的过程:第一步,Servlet 接收前端发来的 HTTP 请求;第二步,对请求进行处理;第三步,将处理结果返回给前端系统。这个过程非常清晰明了。

但是,这里存在一个潜在的问题。如第 3 章所述,当客户端请求 Servlet 提供服务时,容器(例如 Tomcat)会为每个请求分配一个线程,并在这个线程中运行 Servlet 代码,进而为请求提供服务。设想,如果对某个请求的处理需要很长时间才能完成,那么,这个线程会一直被这个请求所占用。由于容器中能够分配的线程数是有限的,因此,如果存在过多的需要长时间才能完成的请求,可能会导致容器的线程资源被全部占用,进而无法为后续请求提供服务。那么如何解决这个问题呢?基于 AsyncContext 的异步请求处理技术就是解决这个问题的利器。

12.1.1 AsyncContext 入门示例

先看一个简单的利用 AsyncContext 对请求进行异步处理的例子。在这个例子中,通过延时来模拟一个需要较长时间才能完成的业务逻辑处理,从而对 AsyncContext 的使用有一个初步了解。为此,在 IDEA 中新建一个名为 ch12 的 Java Web 工程,新建完成的工程如图 12-1 所示。

在 ch12 工程下,新建名为 com.ttt.servlet 的程序包,并在其下新建名为 FirstAsyncServlet 的 Servlet 程序。FirstAsyncServlet.java 的代码如下:

```
package com.ttt.servlet;
import jakarta.servlet.AsyncContext;
import jakarta.servlet.ServletException;
import jakarta.servlet.annotation.WebServlet;
```

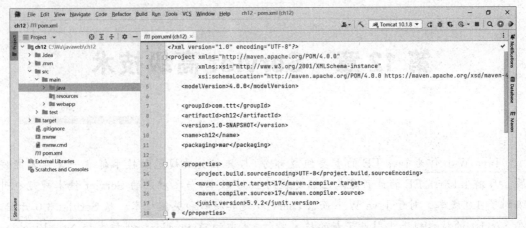

图 12-1 新建的名为 ch12 的 Java Web 工程

```java
import jakarta.servlet.http.HttpServlet;
import jakarta.servlet.http.HttpServletRequest;
import jakarta.servlet.http.HttpServletResponse;
import java.io.IOException;
import java.io.PrintWriter;
import java.text.SimpleDateFormat;
import java.util.Date;
@WebServlet(
    asyncSupported = true,
    urlPatterns = { "/fa" },
    name = "FirstAsyncServlet"
)
public class FirstAsyncServlet extends HttpServlet{
    private final SimpleDateFormat sdf = new SimpleDateFormat("yyyy-MM-dd HH:mm:ss");
    @Override
    public void doGet(HttpServletRequest request, HttpServletResponse response)
            throws ServletException, IOException{
        response.setContentType("text/html;charset=UTF-8");
        PrintWriter out = response.getWriter();
        out.println("访问 Servlet 开始时间: " + sdf.format(new Date()) + " <br/> ");
        out.flush();
        AsyncContext asyncContext = request.startAsync(request,response);
        asyncContext.setTimeout(60000);   //设置最大的超时时间,单位是毫秒
        new Thread(new MyWorker(asyncContext)).start();
        out.println("访问 Servlet 结束时间: " + sdf.format(new Date())+ " <br/> ");
        out.flush();
    }
    private class MyWorker implements Runnable{
        private AsyncContext actx = null;
        public MyWorker(AsyncContext actx){
            this.actx = actx;
        }
        @Override
```

```java
        public void run(){
            try{
                //延迟 10 秒,模拟业务处理时间消耗
                Thread.sleep(10000);
                PrintWriter out = actx.getResponse().getWriter();
                out.println("业务处理结束时间: " + sdf.format(new Date()) +
                    " <br/> ");
                out.flush();
                actx.complete();
            } catch (Exception e){
                e.printStackTrace();
            }
        }
    }
}
```

在这个 Servlet 程序中,使用如下语句:

```java
@WebServlet(
    asyncSupported = true,
    urlPatterns = { "/fa" },
    name = "FirstAsyncServlet"
)
```

这段代码告知容器:这个 Servlet 使用 AsyncContext 异步对象对请求的处理做延时响应,也就是说,这个 Servlet 对客户端的请求处理时间可能比较长,容器可以释放分配给这个 Servlet 的线程对象。然后,在对请求的处理代码中,使用如下语句:

```java
AsyncContext asyncContext = request.startAsync(request,response);
asyncContext.setTimeout(60000);    //设置最大的超时时间,单位是毫秒
new Thread(new MyWorker(asyncContext)).start();
```

创建 asyncContext 对象,并设置处理请求的最长时间时 60000 毫秒,然后,创建新的 MyWorker 线程,并在 MyWorker 线程中对请求进行处理。在 MyWorker 线程中,使用如下语句:

```java
Thread.sleep(10000);
PrintWriter out = actx.getResponse().getWriter();
out.println("业务处理结束时间: " + sdf.format(new Date()) + " <br/> ");
out.flush();
actx.complete();
```

首先延时 10 秒来模拟处理请求需要的时间。延时 10 秒后,再向客户端发送当前的日期时间数据信息,最后使用 asyncContext 对象的 complete()方法告知容器已经完成对本次请求的处理。注意,当线程完成对请求的处理后,务必使用 asyncContext 对象的 complete()方法告知容器已经完成对本次请求的处理。

现在运行这个程序,再浏览器的地址栏输入"/fa",将显示如图 12-2 所示的界面。

在图 12-2 的界面中,注意箭头 2 所指示的表示浏览器再等待后台服务程序响应的标识:在发送请求的 10 秒内,这个标识一直在转动,表示浏览器在等待后台服务程序的响应。那么,为什么是在 10 秒之内呢?因为在后台服务程序的处理代码中使用了 sleep()语句模

拟了 10 秒的处理时间。10 秒后，浏览器将显示如图 12-3 所示的结果。

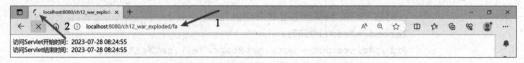

图 12-2　对请求的异步处理运行结果

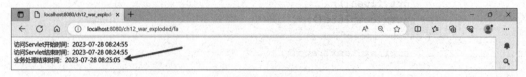

图 12-3　异步处理执行结束后的结果

从图 12-3 可以看出，10 秒后显示了异步线程发送到浏览器的结果；观察一下异步线程共发送到浏览器的时间，与接收到请求的时间正好相差 10 秒。这个 10 秒正是模拟请求处理所需要的时间。

12.1.2　AsyncContext 接口

从 12.1.1 小节可以看出，AsyncContext 接口在异步处理请求中扮演着重要角色。从 Servlet 3.0 开始，调用 HttpServletRequest 对象的 startAsync()方法即可创建 AsyncContext 接口对象，进而可以采用异步技术对 HTTP 请求进行处理。AsyncContext 接口对象的常用方法及其描述如表 12-1 所示。

表 12-1　AsyncContext 接口对象的常用方法及其描述

序号	方法名	描述
1	void addListener(AsyncListener listener)	在 AsyncContext 对象上设置一个异步过程监听器来监听 AsyncContext 对象的状态变化
2	void complete()	当 AsyncContext 对象完成异步任务时，务必调用这个方法通知容器，以便容器可以执行关闭 AsyncContext 对象关联的请求对象和响应对象
3	void dispatch(String path)	将与 AsyncContext 对象关联的请求对象和响应对象转发到指定的 path 进行处理
4	ServletRequest getRequest()	获取与 AsyncContext 对象关联的请求对象
5	ServletResponse getResponse()	获取与 AsyncContext 对象关联的响应对象
6	long getTimeout()	获取与 AsyncContext 对象关联的超时设置，这个超时设置是通过 AsyncContext 对象的 setTimeout()方法设置的
7	void setTimeout(long timeout)	设置 AsyncContext 对象处理请求的最长时间

12.1.3　AsyncListener 监听器接口

在 AsyncContext 接口对象中，可以通过调用 addListener()方法设置实现 AsyncListener

接口的监听器来监听 AsyncContext 对象的状态变化。AsyncListener 接口的方法及其描述如表 12-2 所示。

表 12-2　AsyncListener 接口的方法及其描述

序号	方法名	描述
1	void onComplete(AsyncEvent event)	当 AsyncContext 对象的 complete()方法被调用时,容器将调用该方法通知异步操作已经正常结束
2	void onTimeout(AsyncEvent event)	当异步操作的执行时间超过所设置的最长执行时间,容器将调用这个房通知异步操作超时
3	void onError(AsyncEvent event)	当不能正常完成异步操作时,容器将调用这个方法通知异步操作出现了错误
4	void onStartAsync(AsyncEvent event)	当再次 HttpServletRequest 对象的 startAsync()方法启动一个新的异步操作时容器将调用这个方法

下面举例说明如何监听 AsyncContext 对象的状态变化。在这个名为 SecondAsyncServlet 的 Servlet 例子中,首先调用 HttpServletRequest 对象的 startAsync()方法得到 AsyncContext 对象,然后,在这个异步对象上设置一个 AsyncListener 监听器,并在监听器中显示 AsyncContext 对象的状态信息。SecondAsyncServlet.java 的代码如下：

```java
package com.ttt.servlet;
import jakarta.servlet.AsyncContext;
import jakarta.servlet.AsyncEvent;
import jakarta.servlet.AsyncListener;
import jakarta.servlet.ServletException;
import jakarta.servlet.annotation.WebServlet;
import jakarta.servlet.http.HttpServlet;
import jakarta.servlet.http.HttpServletRequest;
import jakarta.servlet.http.HttpServletResponse;
import java.io.IOException;
import java.io.PrintWriter;
import java.text.SimpleDateFormat;
import java.util.Date;
@WebServlet(
    asyncSupported = true,
    urlPatterns = { "/sa" },
    name = "SecondAsyncServlet"
)
public class SecondAsyncServlet extends HttpServlet{
    private final SimpleDateFormat sdf = new SimpleDateFormat("yyyy-MM-dd HH:mm:ss");
    @Override
    protected void doGet(HttpServletRequest request, HttpServletResponse response)
        throws ServletException, IOException{
        response.setContentType("text/html;charset=UTF-8");
        PrintWriter out = response.getWriter();
        out.println("访问 Servlet 开始时间: " + sdf.format(new Date()) + " <br/> ");
        out.flush();
        AsyncContext asyncContext = request.startAsync(request,response);
```

```java
            asyncContext.addListener(new AsyncListener() {
                @Override
                public void onComplete(AsyncEvent asyncEvent) throws IOException{
                    asyncEvent.getSuppliedResponse().getWriter().println("完成异步
                        处理!");
                }
                @Override
                public void onTimeout(AsyncEvent asyncEvent) throws IOException{
                    asyncEvent.getSuppliedResponse().getWriter().println("异步处理
                        超时!");
                }
                @Override
                public void onError(AsyncEvent asyncEvent) throws IOException {
                    asyncEvent.getSuppliedResponse().getWriter().println("异步处理
                        出现错误!");
                }
                @Override
                public void onStartAsync(AsyncEvent asyncEvent) throws IOException{
                    asyncEvent.getSuppliedResponse().getWriter().println("重新初始
                        化了异步处理!");
                }
            });
            asyncContext.setTimeout(60000);   //设置最大的超时时间,单位是毫秒
            new Thread(new MyWorker(asyncContext)).start();
            out.println("访问 Servlet 结束时间: " + sdf.format(new Date())+ " <br/> ");
            out.flush();
    }
    private class MyWorker implements Runnable{
        private AsyncContext actx = null;
        public MyWorker(AsyncContext actx){
            this.actx = actx;
        }
        @Override
        public void run(){
            try{
                //延迟 10 秒,模拟业务处理时间消耗
                Thread.sleep(10000);
                PrintWriter out = actx.getResponse().getWriter();
                out.println("业务处理结束时间: " + sdf.format(new Date()) + " <br/> ");
                out.flush();
                actx.complete();
            } catch (Exception e){
                e.printStackTrace();
            }
        }
    }
}
```

这个 Servlet 程序与 12.1.1 小节的 FirstAsyncServlet 程序类似,只是在获取 AsyncContext 兑现之后,使用如下代码:

```
asyncContext.addListener(new AsyncListener() {
    @Override
    public void onComplete(AsyncEvent asyncEvent) throws IOException{
        asyncEvent.getSuppliedResponse().getWriter().println("完成异步处理!");
    }
    @Override
    public void onTimeout(AsyncEvent asyncEvent) throws IOException{
        asyncEvent.getSuppliedResponse().getWriter().println("异步处理超时!");
    }
    @Override
    public void onError(AsyncEvent asyncEvent) throws IOException{
        asyncEvent.getSuppliedResponse().getWriter().println("异步处理出现
            错误!");
    }
    @Override
    public void onStartAsync(AsyncEvent asyncEvent) throws IOException {
        asyncEvent.getSuppliedResponse().getWriter().println("重新初始化了异步
            处理!");
    }
});
```

以上代码设置了 AsyncContext 对象状态变化的监听器。当 AsyncContext 对象的状态发生变化时,调用相应的回调方法。现在运行这个程序,在浏览器地址栏输入"/sa",显示如图 12-4 所示。

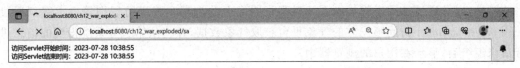

图 12-4　使用 AsyncListener 开始运行的结果

10 秒后,将显示如图 12-5 所示的结果。

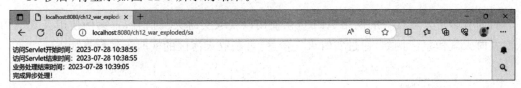

图 12-5　使用 AsyncListener 10 秒后的运行结果

从图 12-4 和图 12-5 可以看出,启动浏览器访问"/sa"也就是 SecondAsyncServlet 服务程序 10 秒前后的结果对比:在 10 秒后,容器会调用 AsyncListener 接口对象的 onComplete()方法告知异步操作执行完毕。

12.2　Non Blocking I/O 技术

为了进一步提升 AsyncContext 异步操作的效率,在 Servlet 3.1 规范中引入了 Non Blocking I/O 机制,并定义两个专门用于该机制的两个接口:ReadListener 监听器接口和

WriteListener 监听器接口。使用这两个接口,可以通过异步方式进一步提升容器线程的使用效率。

Non Blocking I/O 技术

12.3 案例:使用 AsyncContext 访问第三方系统

本案例通过使用 AsyncContext 访问第三方的气象平台获取气象数据,并将从第三方系统得到的信息进行适当组织后,显示在用户的浏览器中。

12.3.1 案例目标

在开发基于 Java Web 的应用系统时,经常需要访问第三方的 Web 系统,从而获得第三方应用系统提供的数据。常见的这样应用包括:通过访问语音识别应用系统进行语音到文字的转换,以及通过访问车牌识别系统对某张图像上的汽车车牌进行识别等。本案例通过使用 AsyncContext 访问第三方的气象平台获取气象数据,并将从第三方系统得到的信息进行适当组织后,显示在用户的浏览器中。

12.3.2 案例分析

目前有多个开放接口的网络 API 可供使用以获取实时天气信息,本案例使用天气 API 获取实时天气信息。例如,通过实时天气信息,可获取天河区的实时天气信息如下:

```
{
    "status": "ok",
    "api_version": "v2.6",
    "api_status": "alpha",
    "lang": "zh_CN",
    "unit": "metric",
    "tzshift": 28800,
    "timezone": "Asia/Shanghai",
    "server_time": 1690785828,
    "location": [
        23.130358,
        113.368509
    ],
    "result": {
        "realtime": {
```

```
        "status": "ok",
        "temperature": 33.94,
        "humidity": 0.48,
        "cloudrate": 0.97,
        "skycon": "CLOUDY",
        "visibility": 15.06,
        "dswrf": 638.9,
        "wind": {
            "speed": 9.79,
            "direction": 134.31
        },
        "pressure": 100148.6,
        "apparent_temperature": 35.9,
        "precipitation": {
            "local": {
                "status": "ok",
                "datasource": "radar",
                "intensity": 0
            },
            "nearest": {
                "status": "ok",
                "distance": 17.51,
                "intensity": 0.1875
            }
        },
        "air_quality": {
            "pm25": 28,
            "pm10": 39,
            "o3": 116,
            "so2": 6,
            "no2": 15,
            "co": 0.5,
            "aqi": {
                "chn": 40,
                "usa": 84
            },
            "description": {
                "chn": "优",
                "usa": "良"
            }
        },
        "life_index": {
            "ultraviolet": {
                "index": 2,
                "desc": "很弱"
            },
            "comfort": {
                "index": 0,
                "desc": "闷热"
            }
        }
```

```
        },
        "primary": 0
    }
}
```

以上代码所获得的结果是一个 JSON 字符串数据,因此,需要在程序中对这个 JSON 数据进行解析后显示在客户端系统中。本案例将结合 Servlet、JSP、JavaScript、HTML、CSS 等进行数据获取及展示,因此,需要设计和编写如下程序文件。

(1) 入口 Servlet:用 Weather.java 文件。该 Servlet 使用 AsyncContext 从彩云气象 API 获取实时天气信息,然后将请求转发给 weather.jsp 进行展示。

(2) 天气信息展示:用 weather.jsp 文件将 Weather 转发的 JSON 天气信息进行解析和展示,还有相关的样式文件 weather.css 及 JavaScript 处理文件 weather.js。

12.3.3 案例实施

1. Weather 服务端代码

首先编写 Weather.java 文件的代码。这个 Servlet 使用 AsyncContext 从彩云气象 API 获取实时天气信息,然后将请求转发给 weather.jsp 进行展示。Weather.java 的代码如下:

```
package com.ttt.servlet;
import jakarta.servlet.AsyncContext;
import jakarta.servlet.RequestDispatcher;
import jakarta.servlet.ServletException;
import jakarta.servlet.annotation.WebServlet;
import jakarta.servlet.http.HttpServlet;
import jakarta.servlet.http.HttpServletRequest;
import jakarta.servlet.http.HttpServletResponse;
import java.io.IOException;
import java.net.URI;
import java.net.http.HttpClient;
import java.net.http.HttpRequest;
import java.net.http.HttpResponse;
import java.time.Duration;
@WebServlet(asyncSupported = true,
        urlPatterns = { "/Weather" },
        name = "WeatherServlet")
public class Weather extends HttpServlet{
    private AsyncContext ac;
    @Override
    protected void doGet(HttpServletRequest req, HttpServletResponse resp)
      throws ServletException, IOException{
        req.setCharacterEncoding("UTF-8");
        ac = req.startAsync(req, resp);
        ac.setTimeout(60000);   //设置超市时间为 60 秒
        Thread t = new Thread(new MyWorker());
        t.start();
    }
```

```
        private class MyWorker implements Runnable {
            @Override
            public void run(){
                HttpClient client = HttpClient.newBuilder()
                        .version(HttpClient.Version.HTTP_1_1)
                        .connectTimeout(Duration.ofSeconds(20))
                        .build();
                HttpRequest request = HttpRequest.newBuilder()
                        .uri(URI.create("https://api.caiyunapp.com/v2.6/" +
                            "TAkhjf8d1nlSlspN/113.368509,23.130358/realtime"))
                        .build();
                HttpResponse<String> response;
                try{
                    response = client.send(request, HttpResponse.BodyHandlers.ofString());
                    ac.getRequest().setAttribute("info", response.body());
                    RequestDispatcher rd = ac.getRequest().
                            getRequestDispatcher("./jsp/weather.jsp");
                    rd.forward(ac.getRequest(), ac.getResponse());
                    ac.complete();
                } catch (IOException | InterruptedException | ServletException e) {
                    throw new RuntimeException(e);
                }
            }
        }
    }
```

2. 编写实时天气信息显示页面

实时天气信息显示页面包括以下文件：weather.jsp 将 Weather 转发的 JSON 天气信息进行解析和展示，还有相关的样式文件 weather.css 及 JavaScript 处理文件 weather.js。

weather.jsp 的代码如下：

```
<%@page contentType="text/html;charset=UTF-8" language="java" %>
<html>
<head>
    <title>实时天气信息</title>
    <meta charset="utf-8">
    <link href="${pageContext.servletContext.contextPath}/css/weather.css"
        rel="stylesheet" type="text/css" />
    <script type="text/javascript"
src="${pageContext.servletContext.contextPath}/js/weather.js"></script>
</head>
<body onload='display(${requestScope.info})'>
    <h2>以下是实时天气信息</h2>
    <div id="weather"></div>
</body>
</html>
```

样式文件 weather.css 的代码如下：

```
body{
    width:100%;
```

```css
    margin: 0 auto;
    text-align: center;
}
p{
    font-size: 20px;
    font-family:楷体, serif;
}
```

JavaScript 处理文件 weather.js 的代码如下：

```javascript
function display(info){
    let weather = document.getElementById("weather");
    let div = document.createElement('div');
    div.className = 'book';
    div.innerHTML =
        '<p>温度：' + info.result.realtime.temperature + '</p>' +
        '<p>湿度：' + info.result.realtime.humidity * 100 + '%' + '</p>' +
        '<p>天气状况：' + info.result.realtime.skycon + '</p>' +
        '<p>能见度：' + info.result.realtime.visibility + '公里' + '</p>' +
        '<p>空气质量,PM2.5：' + info.result.realtime.air_quality.pm25 + '</p>' +
        '<p>空气质量,PM10：' + info.result.realtime.air_quality.pm10 + '</p>' +
        '<p>空气质量,so2：' + info.result.realtime.air_quality.so2 + '</p>';
    weather.appendChild(div);
}
```

现在运行这个程序，在浏览器地址栏输入"/Weather"，显示如图 12-6 所示的实时天气信息页面。

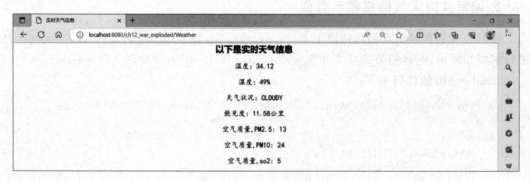

图 12-6 实时天气信息显示结果

从图 12-6 可以看出，程序正确获取了实时天气信息并在浏览器中显示了实时天气信息。

12.4 练习：使用 Thymeleaf 显示气象数据

完成 12.3 节的案例程序功能，但是要求使用 Thymeleaf 作为信息展示方式。也就是说，通过使用 AsyncContext 访问第三方的气象平台获取气象数据，并将从第三方系统得到的信息进行适当组织后显示在用户的浏览器中，但是要求使用 Thymeleaf 作为信息展示方式。

参 考 文 献

[1] 王树生.Java Web开发从0到1[M].北京:清华大学出版社,2023.
[2] 尹有海.Java Web项目开发案例实战[M].北京:中国水利水电出版社,2021.
[3] 张琪.Java Web系统开发与实践[M].上海:上海交通大学出版社,2020.
[4] 高洪岩.Java Web实操[M].北京:电子工业出版社,2021.
[5] 肖海鹏.Java Web应用开发技术[M].北京:清华大学出版社,2020.
[6] 李伟林.从Java到Web程序设计教程[M].北京:电子工业出版社,2019.